NOISE

CONTROL

SOLUTIONS

FOR THE

METAL PRODUCTS

INDUSTRY

By Richard K. Miller, Wayne V. Montone, and Mark D. Oviatt

NOISE CONTROL SOLUTIONS FOR THE
METAL PRODUCTS INDUSTRY

ISBN: 0-89671-000-9

Published by

The Fairmont Press, Inc.
P.O. Box 14227
Atlanta, Georgia 30324

TABLE OF CONTENTS

ABOUT THE AUTHORS

This report was developed by Richard K. Miller & Associates, Inc.,
Consultants in Acoustics, 464 Armour Circle, N.E., Atlanta, Ga.
30324. Established in 1972, the firm is the oldest and largest
in the Southeast providing consulting services exclusively in the
areas of industrial noise control, architectural acoustics, and
environmental noise.

The firm has extensive experience in administering noise control
programs in the metal products industry for clients such as ACF
Industries, Rheem Manufacturing, Houdaille Industries, General
Motors, General Electric, Atlantic Steel, Monroe Auto Equipment,
Abex, Federal Mogul, and Essex.

INTRODUCTION

The approaches to noise control which are presented in this report are designed to address noise problems in the metal products industries in a specific manner. It should be recognized, however, that machinery usage will vary
from plant to plant. While the general approaches to noise reduction presented in this report should be applicable to a wide variety of plants,
careful engineering judgement should be made for each potential application
to insure acoustical, production, and safety constraints are considered and
dealt with.

1 - GENERAL APPROACHES TO NOISE CONTROL

Three approaches to noise control should be considered for any noise problem:

1. The noise *source* may be modified.

2. Noise may be blocked or reduced along the *path* from the source to the receiver.

3. Sound may be isolated from the *receiver* by means of barriers, operator location, or hearing protection.

The optimum approach for any operation must be determined based on acoustical effectiveness, production compatibility and economics. It should be pointed out that OSHA recognizes hearing protective devices as only a temporary solution to noise exposure, and stipulates that other engineering methods must be employed as permanent compliance measures.

The first step in reducing noise is to define specifically how the acoustic energy is being generated. All noise sources generate sound by one of the following two mechanisms:

1. Acoustical radiation from a vibrating surface.

2. Aerodynamic turbulence.

Six types of noise control systems may be considered to solve any noise problem:

1. Sound barriers.

2. Sound absorbers.

3. Vibration damping.

4. Vibration isolation.

5. Mufflers.

6. Machine redesign, process modification, or noise source elimination.

Each of these six conceptual approaches is considered in the noise control solutions for specific items of machinery discussed in this report.

The Williams-Steiger Occupational Safety and Health Act of 1970 (Public Law 91-596) was established "to assure safe and healthful working conditions for working men and women...." The Occupational Safety and Health Administration (OSHA) of the U.S. Department of Labor is delegated the responsibility of implementing and enforcing the law.

Title 29 CFR, Section 1910.95 promulgates regulations for the protection of employees from potentially dangerous noise exposure. A copy of the section is presented in Figure 2.1. Proposed revisions to this regulation were published in the Federal Register of October 24, 1974. This revision is still under consideration.

While the OSHA regulations establish a maximum noise level of 90 dBA for a continuous 8 hour exposure during a working day, higher sound levels are allowed for shorter exposure times. Thus, for cyclic operations, it is necessary to compute the employee's noise dose, or percent allowable exposure for actual operation.

Example:

A machine generates sound levels of 95 dBA for 1 minute during each cycle, 200 times per day. From Figure 2.1, the operator's daily noise dose is:

$$D = \frac{C}{T} = \frac{200 \text{ minutes}}{4 \text{ hours}} = \frac{3.33}{4} = 83\%$$

This dosage is within the OSHA limit of 100%.

§ 1910.95 Occupational noise exposure.

(a) Protection against the effects of noise exposure shall be provided when the sound levels exceed those shown in Table G-16 when measured on the A scale of a standard sound level meter at slow response. When noise levels are determined by octave band analysis, the equivalent A-weighted sound level may be determined as follows:

Figure G-9

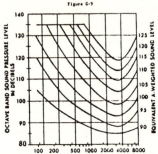

BAND CENTER FREQUENCY IN CYCLES PER SECOND

Equivalent sound level contours. Octave band sound pressure levels may be converted to the equivalent A-weighted sound level by plotting them on this graph and noting the A-weighted sound level corresponding to the point of highest penetration into the sound level contours. This equivalent A-weighted sound level, which may differ from the actual A-weighted sound level of the noise, is used to determine exposure limits from Table G-16.

[1910.95 amended at 39 FR 19468, June 3, 1974]

(b) (1) When employees are subjected to sound exceeding those listed in Table G-16, feasible administrative or engineering controls shall be utilized. If such controls fail to reduce sound levels within the levels of Table G-16, personal protective equipment shall be provided and used to reduce sound levels within the levels of the table.

(2) If the variations in noise level involve maxima at intervals of 1 second or less, it is to be considered continuous.

(3) In all cases where the sound levels exceed the values shown herein, a continuing, effective hearing conservation program shall be administered.

TABLE G-16—PERMISSIBLE NOISE EXPOSURES [1]

Duration per day, hours	Sound level dBA slow response
8	90
6	92
4	95
3	97
2	100
1½	102
1	105
½	110
¼ or less	115

[1] When the daily noise exposure is composed of two or more periods of noise exposure of different levels, their combined effect should be considered, rather than the individual effect of each. If the sum of the following fractions: $C_1/T_1 + C_2/T_2 + \cdots C_n/T_n$ exceeds unity, then, the mixed exposure should be considered to exceed the limit value. Cn indicates the total time of exposure at a specified noise level, and Tn indicates the total time of exposure permitted at that level.

[1910.95 Table G16 amended at 39 FR 19468, June 3, 1974]

Exposure to impulsive or impact noise should not exceed 140 dB peak sound pressure level.

Figure 2.1. OSHA noise regulation.

3 - OVERVIEW OF NOISE PROBLEMS IN THE METALS INDUSTRY

Any acoustical study of steel mills and metal fabrication plants will reveal many of the industries' inherent noise sources to be present in most plants. The noise of these operations can be reduced significantly, and it is these common sources which are addressed in this report.

The following major classes of noise sources are found in most metal industry operations:

- Impact Machines
- Pneumatic Equipment
- Furnaces
- Machine Tools
- Welding
- Material Handling Systems
- Mechanical Equipment

Solution approaches to these types of noise problems are presented in the chapters of this report which follow.

4 - FEASIBILITY

To establish that solution of a noise control problem is feasible, one must consider three areas:

- Acoustical Feasibility
- Production Feasibility
- Economic Feasibility

To establish acoustical feasibility, it must be shown that designs exist which would provide adequate noise reduction.

Each proposed noise control design must be reviewed to insure suitability to the application for which it is intended, and to establish production feasibility. Non-acoustical considerations related to any design include:

a. Employee safety and hygiene.

b. Fire code compliance.

c. Operational integrity:
   1. accessibility to equipment
   2. maintenance serviceability assurance
   3. product quality assurance

d. Machine system compatibility:
   1. mechanical (power, speed, etc.)
   2. service life
   3. ventilation and cooling

Figure 4.1 illustrates the matrix of decisions to be made in determining feasibility. In cases where doubt arises as to acoustical or production feasibility, a design prototype may be required.

The authors of this study, as acoustical consultants, have utilized the following design investigation procedure to establish a basis for acoustical non-feasibility for several industrial operations:

1. A literature search is performed of all available publications in the noise control field and in the general field of the alleged violation.

2. The problem is discussed with colleagues within the professional community to identify where potential solutions to the problem may have been attempted.

3. Recognized authorities in the academic community are solicited for ideas.

4. The literature of all manufacturers of acoustical materials and systems is reviewed for solution approaches, and many are contacted personally.

5. The manufacturer of the noise-producing equipment is contacted, as are several manufacturers of similar equipment.

6. Trade associations are contacted, and the industry-wide state-of-the-art is sought.

7. Solution approaches are solicited from the OSHA personnel involved in the citation.

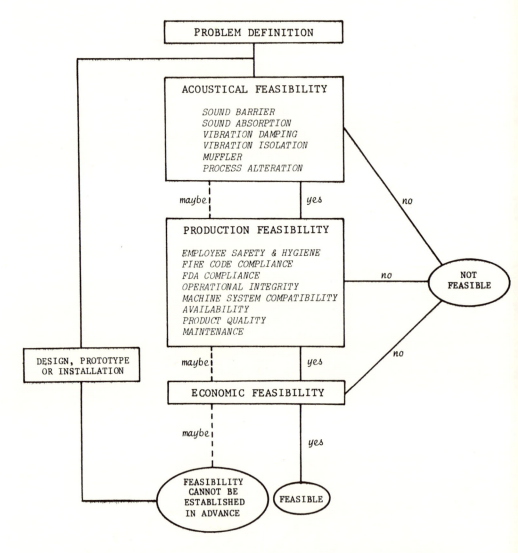

Figure 4.1. Decision matrix for determining noise abatement feasibility.

-8-

5 - LITERATURE SEARCH

A complete search of publications since 1970 revealed numerous articles relating to noise control in metal products plants. The search included review
of periodicals serving both the metals industry and the noise control community. A number of articles related specifically to machinery and operations
covered in various chapters of this report. These articles are referenced
at the end of the appropriate chapters. The following bibliography summarizes additional publications relating to noise control in the metal products
industries.

1. Allen, C. H., "Machinery Noise Reduction by Design," Proceedings of
   Inter Noise 74, pp. 193-196.

2. Allen, C. H., "Reduction of Noise from a Polish and Buff Lathe," Proceedings of Inter Noise 76, pp. 171-174.

3. Cummins, D. P., "Horizontal Boring Machine Noise Reduction," Proceedings of Noisexpo 73, pp. 112-114.

4. Grother, W., "Planning of Noise Control and Results in Some Plants of
   the German Metal Industry," Proceedings of Inter Noise 73, pp. 27-36.

5. Hofmeister, J. R., "Engineering Solutions for Noise Control at GM,"
   Sound and Vibration, Vol. 9, No. 5, May, 1975, p. 39.

6. Kihlman, T., et. al., "Noise Studies in Ship Building," Proceedings of
   Inter Noise 73, pp. 57-61.

7. Kline, E. W., "Truck Factury Noise Control - A Better Way," Proceedings
   of Noisexpo 74, pp. 231-232.

8. Miller, R. K, and Oviatt, M. D., "Noise Control of Rail Car Manufacturing Processes," Proceedings of Noisexpo 77.

9. Miller, T. D., "Machine Noise Analysis and Reduction," *Sound and Vibration*, Vol. 1, No. 3, March, 1967, p. 8.

10. Montone, W. V., "Fire Resistance Codes and Ratings," Proceedings of
    Noisexpo 76, p. 231.

11. Montone, W. V., "Noise Maintenance," Proceedings of Noisexpo 77.

12. Storment, J. W., and Pelton, H. K., "A Practical Approach to Noise Con-
    trol in a Large Manufacturing Plant," Proceedings of Noisexpo 75, pp.
    207-218.

13. Yerges, Lyle F., "Noise Reduction in Metal Cutting Operations," Proceed-
    ings of Noisexpo 74, pp. 227-230.

6 - ECONOMIC ANALYSIS

An economic impact study prepared by Bolt, Beranek and Newman and released
by the U.S. Department of Labor on 17 June 1976 estimated the following for
the primary metals industries (SIC 33):

Number of production workers:	943,800
Workers exposed above 85 dBA:	594,600
Percent workers above 85 dBA:	63%
Compliance cost, 90 dBA:	$1,395,000,000
Compliance cost, 85 dBA:	$2,925,000,000
Monitoring costs (annual):	$11,300,000
Audiometric costs, 85 dBA (annual):	$11,900,000

The following estimates were reported for fabricated metal products manufacturers (SIC 34):

Number of production workers:	997,800
Workers exposed above 85 dBA:	339,300
Percent workers above 85 dBA:	34%
Compliance cost, 90 dBA:	$1,305,000,000
Compliance cost, 85 dBA:	$1,560,000,000
Monitoring costs (annual):	$12,000,000
Audiometric costs, 85 dBA (annual):	$6,800,000

The following were reported for machinery manufacturers (SIC 35):

Number of production workers:	1,383,300
Workers exposed above 85 dBA:	359,700
Percent workers above 85 dBA:	26%
Compliance cost, 90 dBA:	$2,185,000,000
Compliance cost, 85 dBA:	$2,820,000,000
Monitoring costs (annual):	$16,600,000
Audiometric costs, 85 dBA (annual):	$7,200,000

The following were reported by transportation equipment manufacturers (SIC
37):

Number of production workers:	1,133,800
Workers exposed above 85 dBA:	260,800
Percent workers above 85 dBA:	23%
Compliance cost, 90 dBA:	$670,000,000
Compliance cost, 85 dBA:	$1,050,000,000
Monitoring costs (annual):	$13,600,000
Audiometric costs, 85 dBA (annual):	$5,200,000

The publication of this economic impact study met with considerable industry-
wide criticism. To date more accurate cost estimates have not been recognized, however. Due to the controversies which have arisen, it is suggested
that the above costs be considered only an "order of magnitude" estimate.

7 - PRESSES

In terms of the work force exposed to press noise and the difficulty of
silencing, presses probably could be ranked as the number one noise problem
in the United States today.

Presses may be classified into two types:

- Automatic
- Manual

Typically, manual presses operate at speeds of 20-40 strokes per minute,
while the automatic presses may operate up to 1500 strokes per minute. Capa-
cities may range up to 1600 tons. Some presses are horizontal or inclined,
but the majority are vertical. They use single-step dies or progressive dies
and process metals from steel to aluminum to brass of various thicknesses.

Sound levels of press operations may be generated by various mechanisms, and
may be classified as follows:

1. Vibration of the press structures,
   induced by the impact forces.

2. Mechanism noise (clutches, gears,
   etc.).

3. Material handling (ejectors, conveyors, etc.).

The particular noise problem of any press will depend upon its design, speed,
and metal cutting or forming operation. The mechanisms of noise generation
of any specific press may generally be identified by a brief visual auditory
inspection. In some cases, detailed acoustical analysis may be required.
The following sections present approaches to virtually every press noise
problem which may be encountered and is based on information compiled from
a complete literature search and our firm's experience with noise abatement
programs involving several thousand presses.

TYPE OF PRESS

Open back inclined, OBI, presses are generally less rigid than straight side
presses, and are therefore inherently noisier. OBI presses utilize tie rods
to increase rigidity; these must be well maintained to provide minimum noise
levels. Also, it is often found that vibrations induced into the tie rods
radiate considerable noise and require vibration damping or lagging.

CAPACITY

The noise of a press is generally higher as the actual load approaches the
normal capacity of the press. Presses should therefore be selected with 50%
to 100% excess capacity.[1]

DIE DESIGN

In operations involving shearing, blanking and punching using punch presses, the large impact forces exerted by the descending punch on the plate placed upon the die and the shearing action take place simultaneously. A typical operation is illustrated schematically in Figure 7.1. A plate to be blanked is placed upon a stationary die and is engaged by a descending punch whose width is slightly less than the clearance between these spaced parts of the die. The portion of the plate between these spaced parts of the die is, thus, sheared from the body of the plate and moved downward ahead of the punch.[2]

If the punch is designed as illustrated in Figure 7.1(a), all the shearing action occurs simultaneously. A large force is thus involved, as indicated by the height of the shaded area at the right of the figure. If the lower face of the punch is slightly inclined, as illustrated in Figure 7.1(b), the shearing action is distributed over a greater part of the stroke. The maximum force then endures for a shorter period and the duration of the force is increased, as shown by the shape of the shaded area. A further increase in the angle of the lower face of the punch, as shown in Figure 7.1(c), brings about a pronounced decrease in the duration of the force.[2]

In an operation involving the punching of several holes at one stroke of a press, a similar reduction of total force can be attained by stepped punches, as illustrated in Figure 7.2. The punching of successive holes occurs progressively, and not all the punches are operative at a given time. Where pierced blanks are required, punching and blanking operations frequently are performed in multistage dies. Perforations are first made in the material; the material is then moved the length of one stage and the piece is blanked around the perforations while first-stage operations are being performed on the following increment of material. It is desirable, where possible, to equalize the work between stages and to keep the strokes in the several stages out of phase to maintain the individual impulses as small as possible.[2]

Shinaishin[3] reported an 8 dBA reduction with a slanted die, as shown in Figure 7.3.

While slanted or inclined dies and stepped punches are used in many plants, it should be pointed out that this is generally not a complete solution to noise control. Even with these die modifications, sound levels may be above 90 dBA. Also, such die modifications are not applicable to all parts. As pointed out in a recent NIOSH report, "....any improvements in noise level will come by experiment and testing results."[4]

STOCK MATERIAL

Harder materials require greater force, thus producing higher noise levels. Metal working operations involving stainless steel are noisier than those involving cast steel; operations on brass and aluminum are relatively quiet.[5] Shinaishin reported a 14 dB reduction with an experimental substitution of work stock material from steel to a lead-steel composition.[3]

STRUCTURAL VIBRATION

The sound level radiated by a structure is proportional to the vibration level induced in the structure and its radiating surface area. In general, heavy structural members transmit vibration, while lightweight structural elements often vibrate excessively and generate noise. Thus, large lightweight parts such as flywheel guards, frames, and legs are often the major noise problems of a press.

The first approach to vibration analysis is the identification of the specific structural elements generating noise. Both octave band sound pressure levels in the vicinity of the press and octave band vibration levels of the press structural elements must be measured. Then:

1. Peak frequencies in the sound pressure level spectra may be correlated to peak vibration frequencies of the press structure.

2. Measured vibration levels may be correlated to expected (theoretical) sound pressure levels.

The (theoretical) near field sound pressure level above the coincidence frequency of a structure may be computed from the measured or predicted vibration levels by the following relationship:

$$L_p = L_v - 20 \log f + 150$$

where: L_p = sound pressure level, dB
L_v = vibration level, dB re 1.0 g
f = frequency, Hz

The possible use of vibration damping treatment for press noise reduction was investigated by Stewart, Bailey and Daggerhart.[6] Their investigation concluded the following:

For a single degree-of-freedom system, it can be shown that when the ratio of the pulse duration to the system natural period is much less than one, the maximum response can be reduced by increasing the mass of the structure. On the other hand, when the ratio of the pulse duration to the system natural period is much greater than one, i.e., the force is applied slowly, the maximum response occurs while the force is acting. In the latter case, the response is inversely proportional to stiffness, i.e., increasing the stiffness should reduce response and, hence, reduce noise. When the duration of the force is equal to one-half the natural period of the system, a pseudoresonance exists. Control of resonant response can be achieved by detuning the system or adding damping. As pointed out by Harris and Crede[7], however, a tenfold increase in the fraction of critical damping produces a decrease in maximum response of only about nine percent. ....[Thus], damping has little potential for reducing punch press noise.

This limitation applies only to heavy press structural elements. Damping, however, is effective when resonant vibrations are identified in lightweight machine components, such as flywheel guards.

PARTS EJECTION

Air ejection systems are commonly used to eject small parts or scraps from press dies and are sources of high noise levels. Reduction of noise levels can be obtained either by changes in the methods of handling material or by silencing the air system. The following specific approaches may be considered:

1. A thrust silencer may be used as discussed in Chapter 28.

2. Noise caused by high air velocity can be reduced by decreasing the linear flow velocity by increasing the nozzle opening, for same air mass flow. If the diameter of the nozzle is doubled, in a constant volume velocity system, flow velocity is reduced to one-fourth, and noise is reduced nearly 30 dB (noise of air jet varies approximately as fifth power of velocity). However, thrust would also be reduced to one-fourth of original value. For proper ejection, the nozzle must be aimed more accurately and more efficiently toward the target. If the distance from the nozzle to the target were reduced 50%, a 30% velocity reduction would give the same thrust. Experiments must be conducted to determine the maximum thrust required for minimum noise. A silencer designed to these specifications is shown in Figure 7.4.[4]

3. Air used for parts ejection should be controlled by a reducing valve to minimum pressure, and should be regulated to be on only when required for ejection. These measured should also reduce energy consumption considerably.

4. Make the nozzle an integral part of the die set; better, of the die (mount air coupling in die set; possibly of the quick-acting type). The best solution consists of eliminating nozzles completely. A few strategically drilled holes, connected with a common conduit and ending in a quick-connect coupling ejects the parts practically noiselessly.[1]

5. The turbulence and, consequently, the noise, depend not only on the shape of the nozzle, but also on obstacles in the jet. When a jet hits a slot of the kind that is common in a shearing tool, the noise may increase by more than 10 dB, compared to the noise emitted from the same jet impinging on a flat surface. This noise may be minimized by careful aiming of the air jet.[8]

6. An excellent means of practically noiseless part removal consists of using compressed air for creating vacuum. Commercially available devices give highly satisfactory performances.[1]

7. Even with these air noise reduction measures, sound levels may still exceed 90 dBA. A partial enclosure over the die space may be required for additional noise reduction.

8. For stamping ejection first preference should be given to "push-through" evacuation. The stampings should simply fall on the press-bed (bolster plant) and from there they are pushed out (mechanically), unless the press is inclined or the press-bed and the bolster plate have sufficiently large openings for direct evacuation.[2]

9. If for some important reason (secondary operations, part size and/or configuration, etc.) gravity is impractical (or impossible) then use positive action mechanical knock-outs. These may be driven by linkages, air cylinders, hydraulic cylinders, etc.[1]

10. A mechanical parts ejector is available from Lambda Corporation (see Chapter 48).

STOCK FEED

Stock from metal rolls is often positioned using a clamp stock indexer which induces noise-producing impacts into the stock. The best method for noise control is to replace the impact indexer with a new mechanical roll feed.

As an alternative to noise abatement, the indexer may be enclosed and vibration damping pads may be applied to the affected stock. The pneumatic exhaust of indexers should also be muffled. An acoustic infeed tunnel is another approach.

PARTS CONVEYORS

The noise produced by parts rattling on conveyors exiting the press may be reduced by the application of a vibration damping material to the chute. The design in Figure 7.5 utilizing cardboard provided a 10 dBA noise attenuation. An even greater noise reduction would be achieved with a visco-elastic damping layer.[9]

CLUTCHES AND BREAKS

A manually operated press cycle begins with the release of the break and the engagement of a clutch. In mechanical clutches, this involves the impact of a metal pin.

A recent paper[10] identified clutch noise as a primary contributor to press noise. The pin type clutch of manual presses was found to be loudest, with 124 dBA peak levels and decay times up to 0.18 seconds. In the analysis of one press, the clutch provided 66% of the acoustical energy.

Clutches and breaks should be periodically maintained to insure minimum noise levels. In some cases, the impacts may be damped or cushioned, however, satisfactory designs for non-metallic pin clutches have not been developed.

The use of air clutches, which employ pneumatic mechanisms in lieu of positively acting clutches is the recognized solution to noise control for this problem. Air clutches are also somewhat noisy, but may be quieted by applying a damping to the matching surfaces.

The installation of a barrier, as shown in Figure 7.6, is another solution to clutch noise problems. This approach may also shield the noise from other press mechanisms from the operator.

STRIPPER PLATES

In some presses the metal-to-metal impacts of stripper plates contribute significantly to overall press noise. Where this noise is identified, the plate may be damped or a non-metallic contact surface may be used. This may result in noise reduction of up to 10 dBA.

BLANKING PRESS RAM[*]

In large blanking presses the ram is hollow. The forming die runs in grooves on the side of the press, and completely closes off the end of the hollow ram. There are slots in the ram that are used normally when the press is used in blanking operation to extricate the work from the die. This is similar to removal of a cookie from a cookiecutter. These slots are in the side of the ram (see Figure 7.7). When the press is being used in the forming mode, these slots are not required, and when the die "snaps through," it cuts off the work. This gives rise to high noise levels.

This noise may be abated by simply plugging the slots. The plugs may be removed to reconvert to a blanking operation. It is reported that the noise from one press was reduced from 94 dBA to 88 dBA by this method, using a plywood plate cover with a Neoprene gasket.

VIBRATION MOUNTS

The most common and least effective technique of press noise control is the use of vibration mounts. Vibration isolation will reduce vibration transmitted to the floor; however, floor vibrations seldom are an industrial noise problem. One recent study[11] indicated a 15 dB reduction in the vibration level of press legs; however, it is unlikely that vibrations of other press members would be reduced significantly. It should be pointed out that press mounts do have advantages other than noise reduction, and their use is not discouraged. Large noise reduction should not be expected, however.

AIR DISCHARGES

Air noise due to pneumatic control exhausts may be a major noise source of presses, and may be abated by the techniques described in Chapter 40.

DOUBLE IMPACT[7]

Punch presses frequently create unnecessary noise because of double impact. The metalworking operation is performed as the moving head, generally carrying the punch, approaches its lowermost position. The mass of the moving head tends to accelerate the crank, causing it to get ahead of the flywheel and thereby take up any lost motion in the clutch. One impact results when the punch hits the work, and a second impact occurs in the clutch mechanism an instant later as the flywheel catches up with the crank. A properly equipped punch press is provided with a brake, and air cylinder, or a counterbalance to retard the downward motion of the moving head. By preventing the crank from getting ahead of the flywheel, it is possible to eliminate the second impact. Care in adjusting and maintaining these features of presses will eliminate unnecessary impacts. This double impact also subjects bearings, gears, and clutch parts to extra wear, with a subsequent increase in maintenance and cost.

COUNTERBALANCES

Strasser[1] reported the following:

> At the end of the cutting action, when the slug or the blank snaps out suddenly from the sheet metal, all the clearances in the bearing and the elastic strains in the press body structure are suddenly released, and a high intensity noise is created. This can be reduced, sometimes even totally eliminated, by counterbalances (press or die counterbalances). Simple mechanical devices, or more sophisticated hydraulic or pneumatic devices may be employed for this purpose. Counterbalances are more effective in case of heavy stock; in case of thin stock their adjustment becomes too delicate. Such devices should be installed by the press manufacturer; however, they may be applied also later.[1]

BUMPER BLOCKS

Also reported by Strasser[1]:

> In delicate cutting and forming tools it is customary to provide bumper blocks for limiting positively the shut height of the dies. In order to eliminate the noise created by the impact, put a resilient shock-absorbing plastic ring or disc on the bumper blocks. Best results are obtained with comparatively thin layers, about 1/16 inch. Drastic noise reductions are sometimes achieved in this way. Such bumper blocks are very effective in controlling punch penetration in case of cutting operations. Hold punch penetration to a minimum in order to keep noise intensity within reasonable limits.[1]

GEARS

Gears are frequently identified as major contributors of press noise. Noise abatement techniqeus for gears are discussed in Chapter 30.

GUARDS

Flywheel guards should be constructed of damped metal, or open mesh.

ENCLOSURE

The installation of acoustical enclosures to reduce press noise may be considered. With the noise being radiated from the localized area of the die and plunger, a localized enclosure would provide significant noise reduction. Assuming the use of heavy materials and nearly airtight construction, a noise reduction of 10-15 dBA may be expected. Three typical press enclosure applications are described as follows:

1. A partial enclosure of a press by Storment and Pelton[12], as shown in Figure 7.8, achieved a noise reduction from 108 dBA to 89 dBA.

2. Partial enclosures were constructed to supplement noise reduction achieved by a parts ejection silencer. As shown in Figure 7.9. The enclosures resemble a box shaped around the die with the far side and bottom missing, and also serve as safety guards. Noise reductions from 106 to 85 dBA were achieved with a Summit (large size) press, and from 99.5 to 82.5 dBA for a Benchmaster (small size) press.[4]

3. Allen and Ison[13] reported a partial enclosure of ram, die, in-feed, and ejection on a 50 ton test press. A reduction of 13 dBA was obtained for an enclosure; see Figure 7.10. The model enclosure was made of cardboard, 1/2 lb/sq ft, lined with 1" of polyurethane foam. Later a steel enclosure was installed for durability. Total enclosures may also be used for press noise reduction, as shown in Figures 7.11 and 7.12.

The practical limitations of such enclosures, however, have been recognized by these investigators. Shinaishin[3] states, "Enclosures are good solutions when few machines are to be treated and the spacing allows their use." A NIOSH publication[4] describes an enclosure application for a plant with only five presses. Petrie[14] states that, "If frequent access is not required the simplest solution is the complete enclosure around the source." In general, enclosures are applicable only where a limited number of presses are involved and accessibility constraints are not present.

MULTI-SLIDES

The sound levels of multi-slide machines typically range from 90 to 120 dBA. The most common solution for noise reduction of these machines is acoustical enclosures, such as shown in Figure 7.13. Commercially available enclosure units are available from the machine manufacturer, and from:

U.S. ENGINEERING CO., INC.
Guarding Division
5804 Kilgore Avenue
Box 2331
Muncie, IN 47302

INESCON, INC.
Box 1386
Hudson, OH 44236
(216) 655-2146

ENVIRONMENTAL SERVICES
    AND PRODUCTS
Box 1281
Dayton, OH 45410
(513) 258-2196

The primary noise of most multi-slides may be identified as air used for
parts ejection. Air mufflers may be used for reduction of this noise
(see Chapter 28 and 40). On some machines, noise may also be generated
by a cam follower. This noise may be treated with an impact plate as shown
in Figure 7.14.

FEASIBILITY

It should be recognized that it is not feasible to reduce the noise of all
press operations to below the OSHA limits. The following are examples of
cases where sufficient noise reduction may not be possible.

    a. For manually operated presses where die noise is dominant and a
        die enclosure is not possible due to accessibility requirements.

    b. For press rooms where several presses in an area create ambient
        sound levels above 90 dBA, even with silencing.

    c. For operations involving hundreds of presses where economic hard-
        ships caused by production decreases, accessibility problems,
        and installation of enclosures are prohibitive.

    d. Where visibility of the die is required and splashing of lubricant
        precludes the use of an enclosure.

    e. Presses with lightweight frames which are operated at near their
        minimum design capacity are inherently noisy due to excessive
        structural vibrations. Noise control may often be achievable
        by replacement of the press with a new unit.

References

1. Strasser, F., "Noise Pollution in the Press Shop," *Tooling*, October,
    1975, pp. 17-22.

2. Crede, C.E., "Principles of Vibration Control," Chapter 12, *Handbook of
    Noise Control*, edited by C.M. Harris, McGraw-Hill, 1957, pp. 12-26.

3. Shinaishin, O.A., "On Punch Press Diagnostics and Noise Control," Proceedings of Inter Noise 72, October 1976, pp. 243-248.

4. Salmon, V., el.al., "Industrial Noise Control Manual," NIOSH Technical Information, NEW Publication No. 750183, June, 1975.

5. Crede, C.E., "Control of Impact Noise," *The Acoustical Spectrum*, The University of Michigan Press, Ann Arbor, February, 1952, pp. 117-126.

6. Stewart, N.D., Bailey, J.R., and Daggerhart, J.A., "Study of Parameters Influencing Punch Press Noise," *Noise Control Engineering*, Vol. 5, No. 2 September/October, 1975, pp. 80-86.

7. Harris, C.M., and Crede, C.E., *Shock and Vibration Handbook*, McGraw-Hill, 1961, Vol.1.

8. Sahlin, S., and Langhe, R., "Origins of Punch Press and Air Nozzle Noise," *Noise Control Engineering*, November/December, 1974, pp. 4-9.

9. Cudworth, A.L., "Field and Laboratory Examples of Industrial Noise Control," *Noise Control*, Vol. 5, No. 1, 1959, p. 39.

10. Roberts, T.J., Hermann, E.R., and McFee, D.R., "Significance of Punch Press Clutch Noise," presented at 1976 American Industrial Hygiene Conference, Atlanta, May, 1976.

11. Young, R.A., "Effectiveness of Isolators in Reducing Vibration of a 250-Ton Blanking Press," *Pollution Engineering*, Vol.6, No. 12, December, 1974, pp. 32-33.

12. Storment, J.W., and Pelton, H.K., "A Practical Approach to Noise Control in a Large Manufacturing Plant," NOISEXPO 1975, Proceedings pp. 207-218.

13. Allen, C.H., and Ison, R.C., "A Practical Approach to Punch Press Quieting," Proceedings of Inter Noise 73, Copenhagen, pp. 90-94.

14. Petrie, A.M., "Press Noise Reduction," Proceedings of Inter Noise 75, pp. 311-314.

Bibliography

A complete literature search was conducted, and the following bibliography summarizes publications relating to noise control for punch presses:

Allen, C.H., and Ison, R.C., "A Practical Approach to Punch Press Quieting," Proceedings of Inter Noise 73, Copenhagen, pp. 90-94

Allen, C.H., and Ison, R.C., "A Practical Approach to Punch Press Quieting," *Noise Control Engineering*, Vol. 3, No. 1, July/August, 1974, pp.18-23.

Bruce, R.D., "Noise Control for Punch Press," Paper to 21st Annual Technical Conference, American Metal Stamping Association, New York, April, 1970.

Bruce, R.D., "Noise Control of Metal Stamping Operations," *Sound and Vibration*, Vol. 5, No. 11, November, 1971, pp. 41-45.

Bruce, R.D., "A Review of Noise and Vibration Control for Impact Machines," Proceedings of Inter Noise 72, pp. 159-164.

Crede, C.E., "Control of Impact Noise," *The Acoustical Spectrum*, The University of Michigan Press, Ann Arbor, FEbruary, 1952, pp. 117-126.

Crede, C.E., "Principles of Vibration Control," Chapter 12, *Handbook of Noise Control*, edited by C.M. Harris, McGraw-Hill, 1957.

Daggerhart, J.A., and Berger, E., "An Evaluation of Mufflers to Reduce Punch Press Air Exhaust Noise," *Noise Control Engineering*, Vol. 4, No. 3, May/June, 1975, pp. 120-123.

Holub, R., "Controlling Punch Press Noise with Lead/Vinyl Curtains," Proceedings of NOISEXPO 1975, pp. 58-61.

Hoover, R.M., "Noise Levels in Metal Stamping Plants," Paper Presented at 21st Annual Technical Conference, American Metal Stamping Association, New York, April, 1970

Petrie, A.M., "Press Noise Reduction," Proceedings of Inter Noise 75, pp. 311-314.

Roberts, T.J., Hermann, E.R., and McFee, D.R., "Significance of Punch Press Noise," Presented at 1976 American Industrial Hygiene Conference, Atlanta, May, 1976.

Sahlin, S., "Origins of Punch Press Noise," Proceedings of Inter Noise 74, Washington, D.C., pp. 221-224.

Sahlin, S., and Langhe, R., "Origins of Punch Press and Air Nozzle Noise," *Noise Control Engineering*, Vol. 3, No. 3, November/December 1974, pp. 4-9.

Shinaishin, O.A., "Punch Press Noise Outline for Analysis and Reduction," General Electric Research Report No. 72CRD103, March 1972.

Shinaishin, O.A., "On Punch Press Diagnostics and Noise Control," Inter Noise 72 Proceedings, October, 1972, pp. 243-248.

Shinaishin, O.A., "Sources and Control of Noise in Punch Presses," Proceedings of Purdue University Conference on Reduction of Machine Noise, May, 1974, p. 240

Shinaishin, O.A., "Impact-Induced Industrial Noise," *Noise Control Engineering*, Vol. 2, No. 1, Winter, 1974, pp. 30-36.

Stewart, N.D., Daggerhart, J.A., and Bailey, J.R., "Identification and Reduction of Punch Press Noise," Proceedings of Inter Noise 74, p. 225

Stewart, N.D., Bailey, J.R., and Daggerhart, J.A., "Study of Parameters Influencing Punch Press Noise," *Noise Control Engineering*, Vol. 5, No. 2, September/October, 1975, pp. 80-86.

Strasser, F., "Noise Pollution in the Press Shop," *Tooling*, October, 1975, pp. 17-22.

Storment, J.W., and Pelton, H.K., "Practical Noise Control in a Large Manufacturing Plant," *Sound and Vibration*, Vol. 10, No. 5, May, 1975, p. 22.

Wasserman, S., and Mitchell, J., "Noise Control for Punch Presses and Cold Headers," Proceedings of NOISEXPO 73, pp. 168-172.

Young, R.A., "Effectiveness of Isolators in Reducing Vibration of a 250 Ton Blanking Press," *Pollution Engineering*, Vol. 6, No. 12, December, 1974, pp. 32-33.

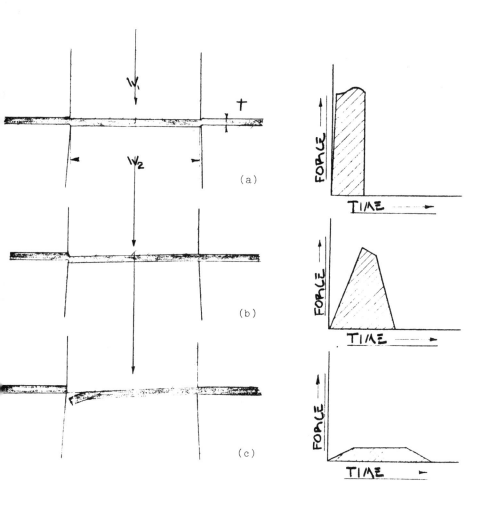

Figure 7.1. Schematic illustration of blanking operation showing the effect of shear angle of the punch. The force-time diagram for each condition is shown at the right.[2]

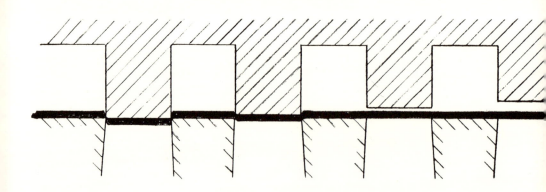

Figure 7.2. Stepped punches for punching several holes at one stroke of a press.[2]

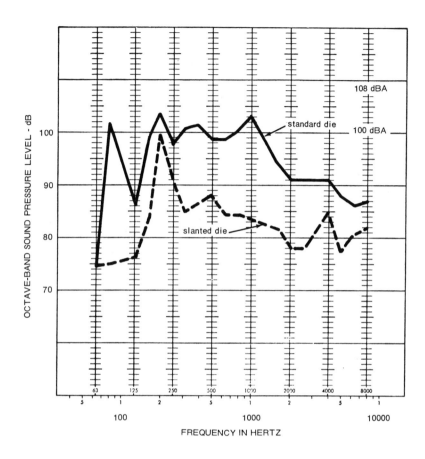

Figure 7.3. Comparison of punch press noise levels: standard die versus slanted die.[3]

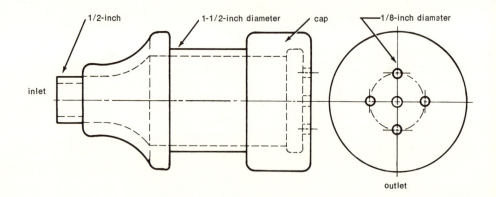

Figure 7.4. Parts ejection nozzle design.

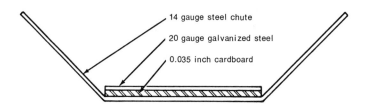

14 gauge steel chute
20 gauge galvanized steel
0.035 inch cardboard

Figure 7.5. Conveyor chute damping.[9]

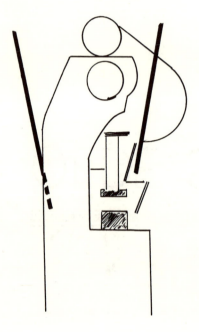

Figure 7.6. Acoustical barriers to shield clutch noise.

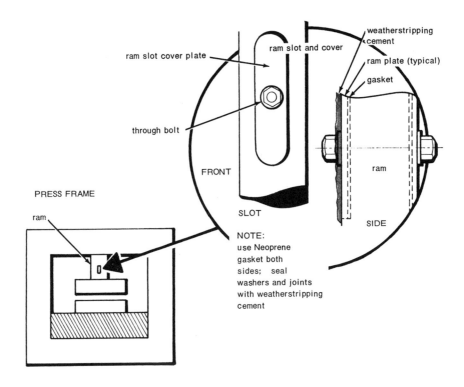

Figure 7.7. Method used to cover slots in blanking press ram.[4]

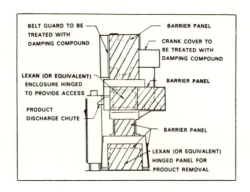

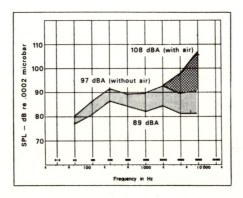

Figure 7.8. Punch press partial enclosure and noise reduction achieved. [12]

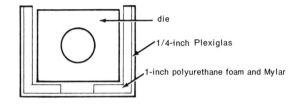

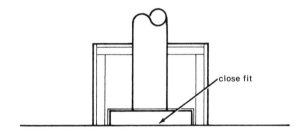

Figure 7.9. Partial enclosure design for press die space.[4]

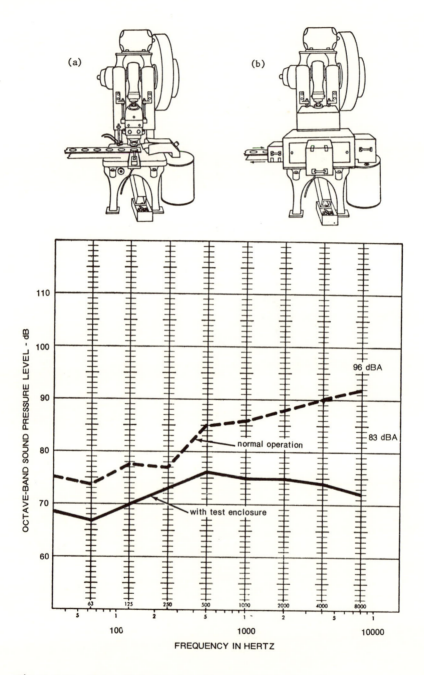

Figure 7.10. Noise 30 inches from 50 ton punch press before (a) and after (b) test cardboard enclosure.[13]

-32-

(Courtesy of Singer Partitions)

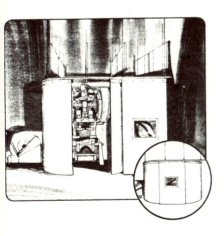 

(Courtesy of Ferro Corporation, Composites Division)

Figure 7.11 Total Enclosures for Punch Presses

Covering of Textolite actually improves appearance of press enclosure.

Construction details are shown in drawing below. Total cost of enclosure was $1.200.

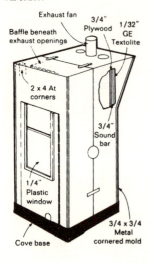

Exhaust fan

3/4" Plywood

1/32" GE Textolite

Baffle beneath exhaust openings

2 x 4 At corners

3/4" Sound bar

1/4" Plastic window

Cove base

3/4 x 3/4 Metal cornered mold

(Courtesy of Soundcoat)

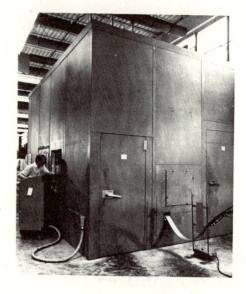

(Courtesy of Industrial Acoustics Company)

Figure 7.12   Total Enclosures for Punch Presses

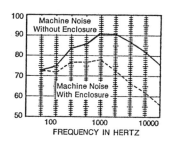

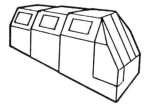

Figure 7.13   Acoustical Enclosure for Multi-Slide.

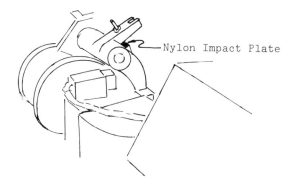

Figure 7.14   Nylon Impact Plate for Cam Follower Mechanism
on Multi-Slide.

8 - PNEUMATIC TOOLS

There are three broad classifications of pneumatic tools: rotary, piston, and percussion. The noise generated by the use of these tools is due to:

- Pneumatic exhaust
- Tool noise: impacts, rotation
- Tool/workpiece interaction

Most of the direct noise from the tool is due to pneumtaic exhaust. The exhaust noise of almost all tools, except for newer models, exceeds 90 dBA under both no-load and load conditions. The pneumatic tool industry was engaged in an increased power per pound race in the early sixties, which lead to an increase in tool noise. This complicates the situation greatly. There are, however, several methods for quieting exhaust noise.

1. Piped away exhaust
2. Expansion volume (muffler)
3. Diffusion at exit
4. Tortuous path

Piped away exhaust systems are by far the most effective where noise is the only consideration. However, due to cost and physical constraints, piped away exhaust systems are generally not the most practical solution to the problem. Piped away systems remove the exhaust from a tool by means of a light-weight hose coupled to the handle. The exhaust air then passes through a muffler or manifold. The major advantage to this system is that it removes exhaust noise to a place remote from the operator and then silences the noise effectively. More than one tool can be connected to a muffler in a piped away exhaust system.

The second solution to tool exhaust noise is the use of an expansion volume, or muffler. This involves the use of an expansion chamber around the exhaust port of the tool to reduce the noise. The system works well, but does have several drawbacks. The first is that mufflers increase tool size and weight. Mufflers also require maintenance to keep their exhaust ports open and free.

The third and fourth solutions to the exhaust noise problem are usually used in conjunction. Diffusion at the exit reduces the exhaust velocity, which, in turn, reduces the noise. This is accomplished by using meshes, sintered metals, felts, and open-cell plastic foam. The major problem with this solution is that clogging occurs at the exhaust port and a pressure drop of 4-8 psi occurs at the exit. The tortuous path method involves sending the exhaust air in a tortuous path through the tool to the exhaust port. This also lowers exhaust velocity and noise.

Tool mechanism noise, other than exhaust noise, is another major noise problem. The driving mechanism in a pneumatic tool creates a large amount of acoustical and vibrational energy. Most of this sound is shielded by the tool itself. Noise from rotary tools using vane-type motors is usually present in the exhaust.

The easiest method to lower the noise level of pneumatic tools is to buy new tools. However, many plants are presented with the problem of tools that work well except that the original exhaust ports have deteriorated. A 10 dBA reduction may be achieved by placing perforated sheet metal over the open exhaust port of vertical grinders where the port is badly damaged. An additional 2 dBA reduction may be obtained by making the thickness of the perforated metal equal to the diameter of the holes.

Pin grinders are very similar to vertical grinders in their noise generation. A basic diffuser can be made for pin grinders by placing 10 ppi foam in the exhaust ports.

Percussive tools, however, present another problem. There are two types of percussive pneumatic tools. One type used a free reciprocating piston to strike repeated blows on a tool held in the end of the barrel. The other type reciprocates a captive piston to pound or ram. Essentialy, both mechanisms have a device which strikes a chisel. The chisel, after being struck, then radiates cylindrical sound waves. An index of different types of chisels appears in Figure 8.1 and their associated noise levels appear in Table 8.1. This data was taken in accordance with CAGI-PNEUROP Test Code for the measurement of sound from pneumatic equipment (ANSI-S.1 1971). From this it is seen that the sound from a pneumatic tool can be decreased slightly by proper chisel selection.

In a previous study by our firm, sound level measurements were performed during chipping operations with a normal and newly sharpened chisel. A sound level reduction of 5 dBA was measured with the sharpened chisel. The employee noise exposure time can also be reduced by use of sharp chisels, since cleaning will be more efficient.

Two other potential methods for quieting percussive tool noise are to coat the chisel with lead or with a visco-elastic material. Neither of these treatments has been found to hold up under load conditions. Coating the chisel with lead also decreases the tool efficiency, as the mass in increased. Special alloy steel chisels have been tried, but the cost is prohibitive for large-scale use, and significant noise reductions have not been documented.

Vibration damping material placed between the piston and the chisel has had some limited success. Once again, with this application, tool efficiency is lost. Tools with this modification are not commercially available.

In a recent test of a Chicago Monarch 25 slag hammer in free operating mode, three different heads were evaluated with regard to associated sound levels. A weld flux scaler head resulted in a sound level of 95 dBA, a damped chisel resulted in a sound level of 86 dBA, and a cold chisel resulted in a sound level of 98 dBA.

The use of chipping tools appears to be inherently noisy, while it is reported that noise level reductions of up to 20 dBA can be achieved by the substitution of needle scalers for chipping hammers for certain operations. It should be pointed out that all needle scalers are not quiet, and that selection of the proper drive mechanism and needle type must be made in order to achieve minimum noise levels.

Assessment of needle scalers must be made with regard to production compatibility as well as noise reduction. Potential advantages of the use of needle scalers include:

a. Needle scalers are reported to provide better weld finishes (U.S. Navy specs require their use for this reason).

b. Needle scalers provide some stress relieving of the weld seam.

c. Needles do not require sharpening.

d. Needles last longer than chisels when used properly (up to 500 hours).

One reported disadvantage of the needle scaler in the experiments conducted was the failure of the tool air exhaust to blow away powder which accumulates in the weld seams. The front of slag hammers provides this function. To solve this problem, the side ports of needle scalers may be blocked to provide front exhaust from these tools also.

Needle scalers may have the following disadvantages:

a. Employees may show a psychological reluctance to change.

b. Weld cleaning may require more time with needle scalers than with slag hammers (however, there have also been reports that needle scalers are faster).

c. Needle scalers are not applicable for all types of welds.

As with chipping hammer chisel selection, the proper selection of needles for needle scalers may reduce the noise level. In a recent test of three needle scalers during a slagging operation, it was found that up to an 8 dBA reduction in the noise level may result by using 2 mm needles instead of 3 mm needles.

Impact wrenches are not always a major noise problem, although some of the larger models can become a noise hazard. There are five steps to be taken for effective noise control of impact wrenches:

1. Contact the manufacturer of the wrench to find out whether retrofit mufflers are available.

2. Pipe away the exhaust from the auxiliary exhaust port if facilities for piped away exhaust are available.

-38-

3. Insert 10 ppi foam in the exhaust port.

4. One element of impact wrehch tool noise is caused by operator misuse. It is common practice for tool operators to keep a wrench working on a nut or bolt after optimal tightening has been achieved. This practice causes the tool to emit excessive noise. The only solution to this problem is to educate the impact wrench operator not to continue this incorrect mode of operation.

5. If possible, replace old tools with automatic shut-off angle nutrunners.

Air cylinders are simple to silence. Noise is generated by two mechanisms: the exhaust, and the stroke. The exhaust can be muffled or piped away and a pad can be placed on the cylinder to cushion the impact of the return stroke. Several cushioned cylinders are commercially available.

In every plant with a pneumatic air system, the problem of air leaks arises. Air leaks are a problem that requires continuing attention because they can generate noise levels of over 90 dBA as well as wasting expensive compressed air.

In addition to the noise generated by the pneumatic tool exhaust, noise is also generated by vibrations induced into the work piece due to grinding or chipping. This noise may be reduced by application of a vibration damping material to the vibrating surface.

There are two basic configurations for the application of damping treatment to structures to increase the loss factor: free layer damping and con-strained layer damping. Free layer damping, also referred to as extensional or surface damping, is the most commonly and easily applied. Damping treat-ment of this type may be a viscous substance which is applied similarly to an automobile undercoating, or may be in the form of a sheet or tile which is bonded to the structure being damped. Constrained layer damping in-volves sandwiching a layer of viscoelastic material between the structure being damped and an outer constraining layer. This type of damping finds application where structural members are quite thick, or where a large vibra-tion reduction is required.

The selection of vibration damping treatment is dependent upon:

- Material type
- Material thickness
- Panel size
- Required vibration reduction

It should be pointed out that it is not necessary to treat an entire surface area to achieve effective damping.

Vibration damping may be installed on structures being ground or chipped, utilizing the following concepts:

a. Placement on a viscoelastic fixture.

b. Application of a damping panel by means of clamps, magnets, or
    fixturing.

Our analysis has indicated that vibration damping may not provide significant vibration or noise reduction for chipping operations of large parts for
the following reasons:

1. "For a single degree-of-freedom system, it can be shown that when the
    ratio of the pulse duration to the system natural period is much less
    than one, the maximum response can be reduced by increasing the mass of
    the structure. On the other hand, when the ratio of the pulse duration
    to the system natural period is much greater than one, i.e., the force
    is applied slowly, the maximum response occurs while the force is act-
    ing. In the latter case, the response is inversely proportional to
    stiffness, i.e., increasing the stiffness should reduce response and,
    hence, reduce noise. When the duration of the force is equal to one-
    half the natural period of the system, a pseudoresonance exists. Con-
    trol of resonant response can be achieved by detuning the system or ad-
    ding damping. As pointed out by Harris and Crede[3], however, a tenfold
    increase in the fraction of critical damping produces a decrease in
    maximum response on only about nine percent,"[4]

2. The thickness of some parts (above 1/2") would render conventional
    damping ineffective, even if applicable. A constrained layer damping
    would be extremely difficult to conform to the irregular shape of many
    parts.

The technical literature is conspicuous in its absence of published case
studies or analysis of vibration damping as an approach to reducing pneumatic chipping hammer noise. It can only be concluded that this common
method of noise control for many situations simply doesn't work for chip-
ping-induced vibrations on thick-walled structures.

An experiment was conducted at a plant to determine the influence of vibration damping treatment on noise levels generated by the use of a chipping
hammer. Three 1/8" plates of approximately 2 square foot area were treated
with an area of 0%, 20%, and 100% with a trowel-on damping compound. The
sound levels measured during chipping were:

Area Damped	Sound Level
0%	113 dBA
20%	112 dBA
100%	111 dBA

It should be pointed out that the damping compound was not completely dry
on the 100% damped plate, and was of a thickness slightly less than 1/8".
From these results, however, it is not expected that a thicker damping

-40-

treatment would provide singificantly greater noise reduction.

In many operations, castings or other metal products are placed on work-benches where they are chipped or ground. Vibrations are transmitted direct-ly from the castings which are being chipped or ground to the metal workbench top. The tables, being of large surface area and relatively light weight, are excellent radiators of sound.

Sound radiation from the table may be reduced by isolating the workpiece vibrations by means of a rubber or soft plastic lining placed on the sur-face of the table. For effective isolation, any rubber material of thick-ness not less than 3/8" may be used. The primary requirement of the rubber material is wearability. In this regard, either conveyor belting or a wear resistant rubber should be satisfactory.

If the workbench is the sight of a welding/grinding operation and the table is used as a ground, placement of constrained layer damping material on the underside of the work surface should be considered.

A less effective but satisfactory alternative to the use of rubber table tops would be table tops constructed of wood. Where wood top benches are presently in use, their use should be continued.

The highest sound levels incurred during chipping operations are when parts are not supported or in contact with a work surface area. Employees should be instructed that parts should be placed on the work surfaces in a manner to provide maximum surface area support.

The practice of chipping parts on wood horses or on surfaces where support of the part is not allowed should be eliminated. It may be necessary to provide additional workbenches to accomodate all chipping requirements which may occur at any given time.

A series of experiments were conducted to determine the influence on noise generation of the manner in which parts are clamped or supported during the grinding and chipping operations. Parts being ground at grinding booths in a cleaning room were clamped by means of pneumatic vises. It was initially theorized that insertion of a vibration absorbing viscoelastic material or wood between the metal vise and part would result in absorption of a signi-ficant amount of vibrational energy, thereby achieving noise reduction. A series of experiments indicated that this was not the case. Measured sound levels were 1 dBA higher with rubber or wood inserts than with the metal vise alone. It is our opinion that the increased sound levels were due to the decreased rigidity of the workpiece in the semi-rigid mounting.

Experiments related to clamping for the chipping operations also indicated that noise levels were not significantly affected by the method or type of clamping.

Damping of vibrations of large castings while being ground is a surprisingly simple process. By placing the casting in a sandbox, vibrations are damped to a degree and they are not transmitted to the normal work surface.

The following sound levels were reported for chipping on a metal casting in a recent experiment by the U.S. Bureau of Mines:

Casting Placement	*dBA*
On concrete floor	113
On sand bed	105
One-half immersed in sand	101
Immersed to top in sand	98
Under sand	94

A pneumtaic tool can and will make more noise when it is not in proper operating condition. One of the main causes of tool degeneration is water in the air lines. This can be corrected by air dryers and water traps. A regular maintenance program for pneumatic tools should be maintained. All rotary tools should also be checked for proper operating speeds. A tool which is being operated in a higher rpm range than that for which it was designed will create higher noise levels.

To insure minimum operating noise levels of all grinders, we recommend that each grinder be inspected for noise levels whenever it is brought to the maintenance department for repair or inspection. The acoustical inspection should include:

1. Checking the sound level with a sound meter.

2. Checking the rpm.

3. A visual inspection of the muffler or exhaust ports.

If excessive sound levels are observed, the grinder should be repaired, or the exhaust muffler should be replaced.

The filter-lubricator is probably the most important component of a pneumatic system. It is the last defense against water and particulate matter. The lubrication added will lower the sound level of a tool by 5 dBA or more and, as a side benefit, greatly increase the life of the tool.

References

1. Anderson, Carl G., "Design Principles for Low-Noise Portable Pneumatic Equipment," 1973 Design Engineering Conference, Pniladelphia, Pennsylvania, April, 1973.

2. Gibbs, C. W., ed., *Compressed Air & Gas Data*, Ingersoll-Rand Corp., Phillipsburg, New Jersey, Second Edition, 1971.

3. Harris, C. M., and Crede, C. E., *Shock and Vibration Handbook*, Vol. 2, McGraw-Hill, New York, 1961.

4. Stewart, N. D., Bailey, J. R., and Daggerhart, J. A., "Study of Parameters Influencing Punch Press Noise," *Noise Control Engineering*, September-October, 1975, pp. 80-86.

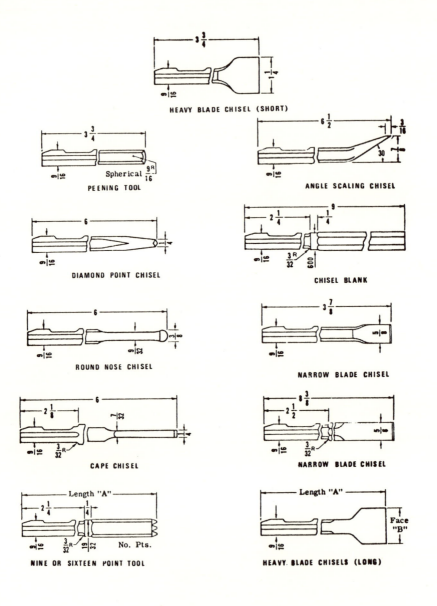

HEAVY BLADE CHISEL (SHORT)

PEENING TOOL

ANGLE SCALING CHISEL

DIAMOND POINT CHISEL

CHISEL BLANK

ROUND NOSE CHISEL

NARROW BLADE CHISEL

CAPE CHISEL

NARROW BLADE CHISEL

NINE OR SIXTEEN POINT TOOL

HEAVY BLADE CHISELS (LONG)

Figure 8.1. Types of chisels used for sound level experiment. All units 1/2" quarter octagon shank.

TABLE I

SOUND LEVELS OF VARIOUS CHISEL TYPES*

Description	Length, in	Sound Level, dBA
Chisel Blank	9	100
Heavy Blade Chisel	3-3/4	100
16 Point Tool	3-3/4	100
Peening Tool	3-3/4	100
Round Nose Chisel	6	97
Daimond Point Chisel	6	99
Cape Chisel	6	98
Narrow Blade Chisel	8-3/8	98
9 Point Tool	8-3/8	101
Narrow Blade Chisel	3-7/8	98
Heavy Blade Chisel	9	97
Heavy Blade Chisel	12	101
Heavy Blade Chisel	18	102
Angle Scaling Chisel	6-1/2	101

* Data obtained from Dr. Carl G. Anderson, Tool Division, Chicago Pneumatic
Tool Company, 2200 Bleecker Street, Utica, New York 13503.

9 - STEEL MILLS

The following sound levels are typical of steel mills[1,2]:

Operation	Sound Level, dBA
Coke Ovens	
Coal breaker	86–94
Larry car operator	82–102
Sinter Plants	
Coke & ore screen	80–95
Burner	82–100
Car shaker	110–120
Blast Furnaces	
Transfer cars	76–104
Furnace building	77–95
Basic Oxygen Furnaces	
Charging floor	81–99
Lancing combustion	100
Fans	95
Electric Furnaces	84–118
Soaking Pits	82–100
Rolling Mills	
Furnace	82–103
Roughing mill	88–100
Hot bed	82–106
Finishing mill	90–110
Fans	85–100
Continuous Pickler	
Feeder	86–100
Shear	86–104
Welder	85–110
Annealing	86–97
Tinning Lines	
Feeder	83–90
Scrubber	86–88
Continuous Galvanizer	
Coil holder	89–104
Shear	85–102
Burr masher	85–93
Material handling	88–115
Nail Mill	
Cold headers	95–105

The employee noise exposure within a steel mill is generally lower than would be recognized from a brief overview of mill operations. Several factors inherently minimize employee noise exposure:

a. Many employees perform all, or a portion of their tasks from enclosures or booths. While many of these enclosures were installed for other production purposes, they also serve as very effective measures to reduce noise exposure.

b. The primary function of many employees is to correct problems and perform adjustments in the mill operation. Thus, they are not engaged in noise-associated activities for a major portion of their work day.

c. Some work areas complete a day's tasks in less than an eight hour shift, limiting noise exposure significantly. Other employees rotate jobs because of heat exposure.

As an initial noise control effort in any mill, the use of employee enclosures should be considered. Enclosure guidelines are presented in Chapter 43. Approaches to noise reduction for many common noise sources found in steel mills are presented in other chapters of this report:

Noise Problem	*Chapter*
Gas Furnaces	11
Electric Arc Furnaces	12
Cut-Off Saws	24
Gears	30
Man Cooler Fans	36
Ventilation Fans	38
Metal-to-Metal Impact Noise	41

The following sections present noise control approaches for various other mill operations.

COUPLINGS

Steel couplings are located between the motor-gear drive and mill stands to provide torque transfer and to insure system alignment. The mill stands cannot operate with an uncoupled shaft. The couplings are relatively quiet when under load, or with a bar in the mill. Noise is generated by vibrations introduced into the coupling when rattling in the unloaded condition. The A-weighted sound level often exceeds 100 dBA at 3'. The noise level of any coupling system is dependent upon its tightness; however, this is not a feasible parameter when considering noise abatement, since all couplings are periodically tightened, and gradually loosen under operational loads.

The only technology known for reducing the coupling noise is a nylon insert type wobbler box. The manufacturer could not estimate the noise reduction to be expected for these couplings. Cost is about $500 each. The experience of many mills indicates that to use these couplings successfully, everything must be in very good alignment, and maintenance must be excellent. They are easily stripped during a cobble and are sensitive to heat. It is questionable if any mill other than a new facility could operate with anything requiring such close tolerances.

Since the couplings are noisy only under no-load conditions, it has been considered to create a continuous false load condition for the couplings. It is reported that this has been attempted in a steel mill in the northwest United States, and provided good experimental noise reduction. The setup, however, involved a bushing propped up with bricks, and more investigation into this approach is clearly needed.

The use of a constrained layer damping treatment to the coupling box may provide noise reduction; however, we are not aware that this approach has ever been tried. Localized enclosures may also be tried.

FINISHING

The sound levels in finishing areas may range up to 110 dBA, and are due to several specific noise sources, including:

  a. Vibration of the stock on the pull-up rolls.

  b. Unscrambling of the stock from piles.

  c. Vibration of stock being sheared.

  d. Structural vibration of shear due to cutting impact.

  e. Dropping of parts into bin or onto stack.

  f. Stock hitting the metal stops.

Noise problems in finishing areas of steel mills have received considerable industry-wide attention for several years without much success in achieving effective noise reduction. To the best of our knowledge, there are no rolling mills in the United States whose finishing areas have sound levels within the OSHA limits. This conclusion is based on an extensive literature search, and contact with numerous manufacturers and professionals in the field of noise control.

Potential noise control solutions were investigated for each of the noise sources, and our findings will be discussed here.

Since finishing involves operators working directly with the steel product, the use of employee enclosures is not practical. While minor items of noise control are sometimes possible for finishing areas, these would generally have no influence due to the overriding effect of noise from more severe sources. The most dominant noise source is vibration of stock on the pull-up rolls.

Vibration of the stock is the primary source of noise on pull-up roll conveyors. Thus, vibration damping of the rolls would not be effective. The potential design solution of coating the exterior of the rolls with a material to cushion the metal-to-metal impacts between the stock and pull-up rolls was investigated. This material must meet the following requirements:

a. Must have a high modulus of elasticity to provide impact isolation.

b. Must provide high wear characteristics.

Conventional plastic and rubber materials which would be ideal as impact isolators would not withstand the wear requirements. Many mills have maintained an ongoing effort to locate a wear resistant roller coating for several years. None of these efforts has resulted in finding a successful material.

We believe the wear problem is inherent with conventional resilient materials. The angle of incidence of the steel bar upon the pull-up roll ranges between 0° and 20°, as shown in Figure 9.1. A wear curve as shown in Figure 9.2 illustrates that compliant materials cannot be successfully used in this application.

COOLING BEDS

Cooling beds which involve sliding of steel bars across rails can generate noise levels in excess of 110 dBA.

The noise generated by the steel motion on the bed is due to frictional vibrations induced by the microscopic roughness of the bar and cooling bed rails. The frictional vibration forces are directly proportional to the coefficient of friction between the surfaces, and independent of the velocity of motion. The high temperature (over $1000^{\circ}F$) of the bar severely limits potential modification to the rails for the purpose of reducing friction forces. The coefficient of friction between steel surfaces is 0.45 - 0.55. This value may be reduced to 0.03 - 0.1 by application of a lubricant to the metal interface. This has been attempted at one mill and was found to be impractical for the following reasons:

a. The lubricant caught fire.

b. The lubricant wore off very rapidly.

c. The lubricant could not be cleaned off the bar for painting.

Polishing the bed rails to a microfinish would reduce friction forces, and resulting noise levels. It is questionable, however, whether such a finish could be maintained under production conditions for any extended period of time, and feasibility of this approach cannot be established.

In view of the limited acoustical effectiveness and probable production

-49-

interference of other design approaches, the use of employee enclosures is the best approach for noise control of this operation.

References

1. Botsford, J. H., "Noise in Mining and Metals," Seventh Institute on Noise Control Engineering, 1973.

2. "Impact of Noise Control at the Workplace," U.S. Department of Labor, L-73-112, October 29, 1973.

General References

1. Botsford, J. H., "Noise Control in the Steel Industry," *Iron and Steel Engineer*, June, 1962, pp. 90-101.

2. "Noise Standards - 1975, Impact on the Steel Industry," American Iron and Steel Institute, 1975.

3. Martin, M. D., "Noise Control in the Steel Industry," *Steel Times*, November, 1972, pp. 809-811.

4. Handley, J. M., and Schiff, M. I., "Feasible Engineering Noise Control for the Primary Metals Industry," *Iron and Steel Engineer*, March, 1975.

5. Campbell, J. M., and Willis, R. R., "Protection Against Noise," *Journal of the Iron and Steel Institute*, May, 1973, pp. 346-352.

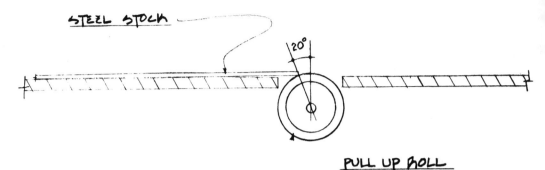

Figure 9.1. Angle of incidence of steel bar impact on pull-up roll.

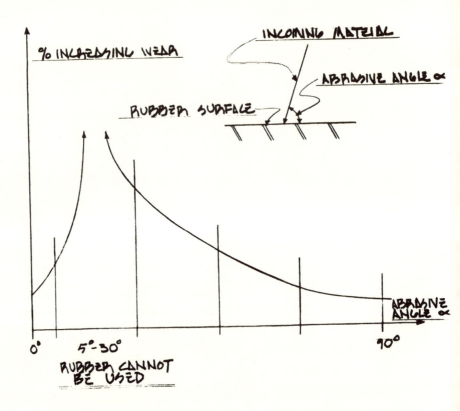

Figure 9.2. Wear curve for wear-resistant rubber grade. The curve is approximate and provides a good general picture of the wear resistance of rubber at different abrasive angles.

With OSHA's National Emphasis Program, noise control in foundries has received more attention than ever. Noise control, however, is not new for foundries; the American Foundrymen's Society developed its first *Foundry Noise Manual* in 1958. With this emphasis, most foundries today are significantly quieter than they were 5-10 years ago, and practical solutions now exist for most foundry noise problems.

Solutions to many noise problems encountered in foundries are presented in other chapters of this report:

Noise Problem	*Chapter*
Pneumatic Tools	8
Gas Furnaces	11
Electric Arc Furnaces	12
Grinding	17
Air carbon Arc Gouging	18
Materials Handling & Vibrators	20
Air Noise	28
Tote Bins	32
Man Cooler Fans	36
Ventilation Fans	37
Metal-to-Metal Impacts	41

The following sections discuss other noise problems found in foundries:

SHAKEOUT

Castings are removed from sand molds by means of a shakeout which dislodges the casting by impacting the flask on a metal deck. Sound levels may range up to 110 dBA.

In automated foundries, it is possible to reduce shakeout noise by means of an acoustical enclosure or partial enclosure. Satisfactory results achieved by such enclosures have been reported by several investigators.[1,2,3] A shakeout enclosure design is shown in Figure 10.1.

In a jobbing shop, where castings of many sizes are handled, it may not be feasible to enclose the shakeout. Two alternative solutions may be considered:

1. Rubber bumper bars (80 durometer, 1" thick and 1½" wide) may be bonded to a ½" metal bar which is spot welded to the shakeout deck. When the shakeout grates vibrate, these bumper blocks repeatedly strike the red hot castings, but they do not stay in contact with them long enough to burn. This design has been successfully employed in foundries, and noise reductions of 7 dB are reported. A service life of up to 9 months has been reported.[1]

2. In another foundry, steel chains with links 1½ inches in diameter were threaded through automobile radiator-hose material. Lengths of chain-threaded hose were placed longitudinally across the shakeout grate and secured with sufficient slack at appropriate intervals to the sides of the shakeout deck. This eliminated the impact of metal against metal, preventing casting damage, and reduced noise. Maintenance costs were not high, and the hose lasted about a month in that particular plant.[3]

CORE OVEN FLAME ADJUSTMENT

One article reports that core oven burners in a foundry were found to vary from 88 to 94 dBA. It was learned that the burners were purposely out of adjustment so that they could be heard. The operators of the ovens were afraid of a flameout, and to assure them of a flame, the operators permitted the burners to run with a low-pitched roar that is characteristic of combustion noise. The solution here was to use an electronic flame detector which will alarm and shut off the gas rather than depending on the noise of the burners to indicate a flame.

CLEANING BOOTHS

It was experimentally determined that operators in small three-sided, open top cleaning (grinding or chipping) booths experience a sound level build-up of approximately 8 dBA due to sound reflection from the booth walls. The sound level during normal grinding operations for an experimental set-up was 99 dBA. This sound level was reduced to 91 dBA by lining the booth walls with a 4" layer of glass fiber thermal insulation.

To provide effective noise reduction within grinding booths, interior wall surfaces may be lined with a 1" minimum thickness of glass fiber or open cell polyurethane foam faced with a protective layer of 24 GA perforated sheet metal with 20% open area, or heavy wire mesh screen.

TUMBLERS

Noise reductions from above 110 dBA to below 90 dBA are reported[1] as a result of lining casting tumblers with resilient wear retardant rubber linings (see Chapter 41 for list of commercially available materials).

References

1. *Foundry Noise Manual*, Second Edition, American Foundrymen's Society, Des Plaines, Illinois, 1966, p. 62.

2. *Control of Noise*, Third Edition, American Foundrymen's Society, Des Plaines, Illinois, 1972, p. 67.

3. Weber, H. J., "Foundry Noise," *Noise Control*, January, 1959, pp. 44-76.

4. Young, R. L., "Practical Examples of Industrial Noise Control," *Noise Control*, Vol. 4, No. 2, March, 1958, p. 11.

General References

1. Huskonen, W. D., "Progress Report on Noise Control," *Foundry*, June, 1973, pp. 42-45.

2. Ihde, W. H., "Divide and Conquer Your Noise Problems," *Foundry*, August, 1973, pp. 61-62.

3. Semling, H. V., "Noise Control and Government Regulation," *Foundry*, February, 1972, pp. 53-55.

4. Warnaka, G. E., "Planning a Quiet Foundry," *Foundry*, May, 1970, pp. 64-71.

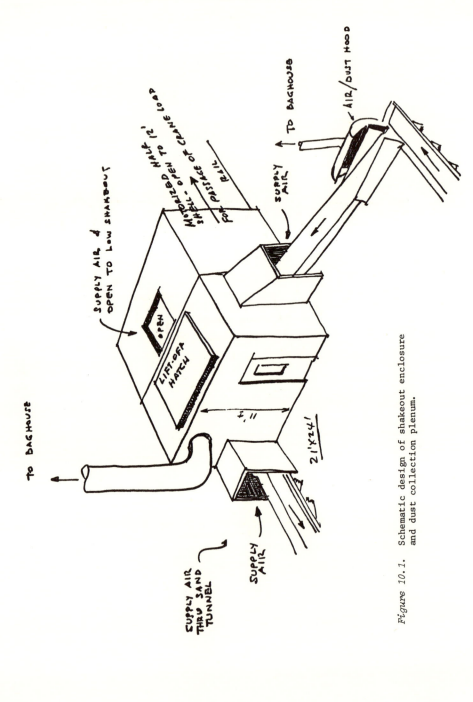

Figure 10.1. Schematic design of shakeout enclosure and dust collection plenum.

11 - GAS FURNACES

Gas furnaces are inherently noisy due to combustion noise. Combustion noise is very similar to air noise from a jet. Combustion creates an additional low frequency component. Whereas jet noise intensity is proportional to approximately 7 times the jet diameter, combustion noise is proportional to approximately 50 times the burner diameter.

Combustion noise is basically derived from variations in the heat release rate. There are two general types of combustion noise, which are:

- Turbulent combustion noise
- Combustion oscillations

Turbulent combustion noise occurs when there are random fluctuations in the rate of combustion in the combustion reaction zone. This type of noise manifests itself as a "roar". Practically all common fuel-burning systems have turbulent reaction zones which produce turbulent combustion noise. For satisfactory combustion, high mean flow velocities with high turbulence intensities are required. For quietness, velocities and turbulence levels should be kept low.

Combustion oscillations are a phenomenon that arises in a small percentage of gas burners. However, when they occur, the oscillations are noticed because of the rumbling, trumpeting or screeching that is emitted by the flame. These oscillations have been studied for over 170 years. In 1896 Rayleigh[3] stated that oscillations will occur when the periodic heat release is in phase with the maximum pressure variations.

Occasionally combustion oscillations will manifest in the form of toroidal vortices. When these vortices interact with the flame front, they give rise to discreet tone generation. A simplified diagram of this phenomenon appears in Figure 11.1.

Due to the complexities of compustion oscillations, there is no definite solution to the problem. The following list of remedies is not necessarily in order of effectiveness, cost, or etc.

1. Altering the air and fuel rates to the burner can often remove oscillations. This is cheap and simple, but frequently either cannot be done at all, or can be done within very narrow limits.

2. Changing the type of burner is frequently effective. A completely different design of burner should be chosen; this modification is rather drastic and often out of the question for political or economic reasons.

3. Altering the configuration of the supply pipework can occasionally solve oscillation problems. This can be done where acoustic standing waves are thought to exist upstream of the burner. This modification is not one which is commonly found to be effective, however.

4. Modifying the combustion chamber geometry can remove oscillations. This can be done by placing a refractory wall part way across the combustion chamber, or by altering the gas flow path (and hence changing the acoustics; an example of this is the partitioning of the first "reversal chamber" in a shell boiler with two combustion chambers exhausting into a common flue). This modification may not be possible for reasons involving heat transfer or static pressure in the system.

5. Providing "pressure release" orifices at various points in the "hot" part of the system. These are often placed near to the burner and need be no more than 1/2 inch in diameter in many cases. Again, these may be impractical because, for instance, of pressure considerations.

6. Modification of the burner head design may alter the flame transfer function so that oscillation disappears. This must, of course, be done so as to retain proper flame stability and combustion quality.

7. Fitting tunable acoustic filters such as quarter-wave tubes or Helmholtz resonators can "tune out" oscillations. These consist simply either of a side-branch tube connected, for instance, to the combustion chamber and fitted with a piston, or a cavity resonator similarly situated, with a variable volume. These are usually successful if fitted in the correct position (near to the burner is often very effective) but can be unsightly. Their chief disadvantage is that they are so sharply tuned that if the frequency of the oscillation wanders (as it frequently does), they go "off resonance" and cease to be effective.

8. A reactive expansion chamber type of flue stack silencer can be effective if the amplitude of oscillation is not too large. This remedy can be quite expensive where a large silencer is required, and in any case does not tackle the problem at its source.

9. In certain situations, it is possible to change the composition of the fuel supplied to the burner. This may be the case where, for instance, a factory manufactures its own gas. The effect of this would be principally to alter the transfer function of the flame. Alternatively, it may be possible to change from liquid to gaseous fuel, or vice versa.

10. Increasing the pressure drop across the burner can prove effective in curing oscillations; this would normally be done in conjunction with a burner head modification. It would, however, usually be accompanied by an increase in the level of broadband turbulent combustion noise, but in many situations this may be unimportant.

11. Altering conditions in the flue stack would change the transfer function of the combustion chamber and may prove to be an effective remedy; the provision of a flue break, for instance, has proven to be successful in the past.

References

1. Cummings, Alan, "Combustion Noise," *BSE*, Vol. 41, September, 1973, pp. 123-131.

2. Briffa, F. E. J., Clark, C. J., and Williams, G. T., "Combustion Noise," *Journal of the Institute of Fuel*, May, 1973, pp. 207-216.

3. Rayleigh, J. W. S., *Theory of Sound*, Vol. II, Second Edition, Dover, New York, 1896, p. 224.

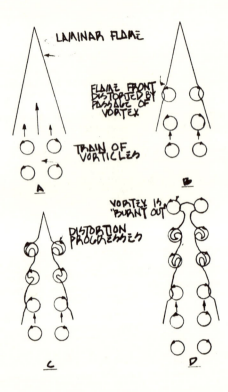

Figure 11.1. Effect of an incident train of toroidal vortices on a laminar flame front.

12 - ELECTRIC ARC FURNACES

Electric arc furnaces are bowl-shaped metal housings with a refractory brick lining in which scrap steel is melted. The melting of the steel is achieved by an arc struck between three graphite electrodes and the steel. The arc potential and current are approximately 450 volts and 60,000 amperes respectively. An internally luminous column is formed between two hot spots, one on the electrode tip and one on the melt. The hot spot and plasma column have temperatures from 3600-4000°C and 5000-6000°C respectively.

The major portion of the electrical energy used in converted into thermal energy which melts steel. Some of the electrical energy is converted into other forms of energy, such as chemical, mechanical, and electromagnetic. These energy forms are manifested by undesirable secondary effects such as exhaust gases, vibration, noise, side wall heating, and light.

Sound levels of up to 115 dBA are generated by the furnace. The noise has two major components, which are electric arc noise, present at all harmonics of 60 Hz, and mechanical noise from scrap movement within the furnace. The noise time history of a melt follows this general pattern:

1. Noisy bore-in.

2. Quiet pool.

3. Noisy melt-down.
   a. Delta cave-ins.
   b. Wall cave-ins.

4. Quiet flat-bath.

Since the furnace does not require constant attention during the melt cycle, employees may usually spend most of the time when high noise levels are being generated in an acoustical enclosure as a means of achieving OSHA compliance. Employees in adjacent areas engaged in other activities, however, may also be exposed to excessive noise levels. The only approach to noise abatement for these employees would be to isolate them by means of a wall or to relocate their work area. This is often not possible, since foundry and steel mill operations often require large work spaces with crane access.

We conclude that there has been no feasible method developed for reduction of the noise from the electric arc furnaces, nor does it appear that any noise control research has ever been undertaken for electric arc furnaces, based on the following:

a. A literature search.

b. Discussions with several furnace users.

c. Information from furnace manufacturers.

Our firm has developed a protocol for such research, and it appears that some noise reduction may be achievable in the future.

Reference

1. Higgs, Roland W., "Sonic Signature Analysis for Arc Furnace Diagnostics and Control," 1974 Ultrasonics Symposium Proceedings, IEEE Cat. No. 74 CHO 896-ISU.

13 - COLD HEADERS

Cold headers constitute the basic machine tool used in the fastener manufacturing industry. Basically, it forms the head on screws, bolts, nails, and rivets.

Lampe described the noise generated by cold headers as follows[1]:

> The impact forces are conducted through the working parts of the machine to the surfaces of the machine and are radiated in the form of airborne noise. Adding to the characteristics of the noise are the changes in acceleration of the ram, which changes the loading of the shaft bearings, meshing of the teeth of the gears, cam peaks, misalignment, knife and gripper contact, and the revolutions of the motor shaft. Vibrations from these sources, which did not already become airborne, are conducted to the surface areas of the castings, the guards, and the doors where they cause additional resonant frequencies and become either airborne or structureborne noise.

The most common approach to noise control for operations involving only a few cold headers is the use of acoustical enclosures. Either partial enclosures for the impact area or complete enclosures may be considered. One enclosure design is discussed in Reference 2. Complete design guidelines for enclosure design are presented in Reference 3.

Cold headers may transmit excessive vibration when mounted on weak floors, which may radiate considerable noise. The sound level of a 300 stroke per minute nail making machine was reduced from 103.5 dBA to 95 dBA[4] by the installation of elastomer type isolators having a static deflection of 0.1" under machine load. This corresponds to a natural period of 100 msec, thus fulfilling the following design goals:

1. The natural period of isolator plus machine should be much greater than the shock pulse duration (10 msec).

2. The natural period of isolator plus machine should be less than the time between pulses (200 msec).

The following redesign approaches may be considered to reduce cold header noise[1]:

1. Use pressing instead of impact forces.

2. Blend the cam curves and reduce peak acceleration.

3. Use resilient material between the die mounting block and the supporting casting surfaces and also between the ram and the header tooling.

4. Use resilient rollers on the cam followers.

5. Place nylon bushings between bearings and castings.

6. Isolate the machinery from the base casting.

7. Isolate the base from the floor.

8. Treat sheet metal guards and doors with damping materials or methods.

9. Use Plexiglas where guarding and visual inspection are necessary.

10. Make the machine easier to remove from the line for inspection, preventive maintenance and repair.

Wolgast reported the reduction of cold header noise using the following approaches:[5]

a. A viscoelastic material was applied to the rocker arms.

b. The flywheel was treated with a vibration damping.

c. The greater part of the noise associated with the tool impacting the billet was identified as being due to the bearing between the crankshaft and the connecting rod. The most promising method of solving this problem appears to be to apply a force, such as from a mechanical or pneumatic spring system, to the cross head block.

References

1. Lampe, Guy, "Cold Header Noise Reduction," NOISEXPO 73 Proceedings, p. 164.

2. Wasserman, S., and Mitchell, J., "Noise Control for Punch Presses and Cold Headers," NOISEXPO 73 Proceedings, pp. 168-172.

3. Miller, R. K., and Montone, W. V., *Handbook of Acoustical Enclosures and Barriers*, The Fairmont Press, 1977.

4. Crocker, M. J., and Hamilton, J. F., "Vibration Isolation for Machine Noise Reduction," *Sound and Vibration*, Vol. 5, No. 11, November, 1971, p. 30.

5. Wolgast, Arne, "Reduction of Noise Emitted from Cold Heading Machines," Inter Noise 76 Proceedings, pp. 473-476.

14 - SCREW MACHINES

Analysis of several screw machines indicated that the following mechanisms are the primary contributors to excessive noise levels:

1. Gears
2. Stock tube vibration

Conventional approaches to gear silencing may be considered to reduce screw machine noise; however, it has been found that this approach requires re-building of existing machines and is uneconomical.[1] The most common method employed for silencing existing machines is the use of acoustical enclosures. While satisfactory sound levels may be achieved with enclosures, two potential disadvantages are apparent:

a. The cost per enclosure ranges from $3000 to $4000, or 10% to 15% of the cost of the machine itself.

b. An enclosure may decrease operator efficiency, resulting in an increased production cost per part.

Considerable design effort has been invested by Davenport, resulting in an enclosure which reduces the noise levels of their machines to 80 dBA.[1,2,3] Their design is reported to minimize adverse production influences.

The manufacturer of Acme-Gridley screw machines has concentrated noise control efforts on the development of quiet gear systems, which are offered on new machines. We have measured the noise levels of these machines to be below 90 dBA. On older Acme-Gridley machines, noise may be reduced by installing removable panels on the front of the machine, as shown in Figure 14.1.

Acoustical enclosures for screw machines are available from the following firms:

Davenport Machine Tool Division
Box 228
Rochester, NY   14601

Keene Corporation
Porta-Fab Division
2319 Grissom Drive
St. Louis, MO   63141

Doug Biron Associates, Inc.
P.O. Box 413
Buford, GA   30518

Robert Sheehan Company
P.O. Box 4544
Anaheim, CA   92803

Singer Partitions, Inc.
444 N. Lakeshore Drive
Chicago, IL 60611

Inescon, Inc.
P.O. Box 1386
Hudson, OH 44236

Noise is generated by the slap of stock against the tube interior. This
noise may be reduced either by enclosing the tubes or by the installation of
tubes with an isolated interior liner.[4,5] Quiet stock feed tubes are available from:

Corlett-Turner Company
9145 King Street
Franklin Park, IL 60131

Frelun Engineering Company
1240 Harrison Avenue
Rockford, IL 61101

The Polymer Corporation
Reading, PA 19603

Referneces

1. Bourne, J. C., "Sound Control Hood for the Davenport Automatic Screw
   Machine, Proceedings of Noisexpo 74, pp. 127-133.

2. Bourne, J. C., "Silencer 2 - Sound Control System for the Davenport
   Automatic," Proceedings of Noisexpo 76, pp. 165-169.

3. Bourne, J. C., "Effectiveness Assessment of the Davenport Silencer 2,
   Proceedings of Noisexpo 77.

4. Cottingham, R. A., and Winnerling, H. A., "An Advanced Low Noise Stock
   Feed Tube," Proceedings of Noisexpo 74, pp. 152-154.

5. Schweitzer, B. V., "Silent Stock Tube for Automatic Screw Machines,
   Noise Control, Vol. 2, No. 2, March, 1956, p. 14.

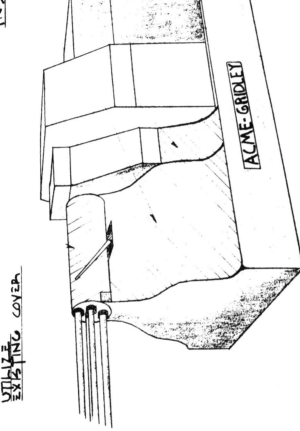

INSTALL PANELS

UTILIZE EXISTING COVER

ACME-GRIDLEY

Figure 14.1. Acoustical enclosure of Acme-Gridley screw machines.

-67-

15 - LATHES AND MILLING MACHINES

Noise may be generated in lathes and milling machines from three sources:

1. Vibration of the workpiece
2. Tool chatter
3. Machine noise (gears, etc.)

One recent study[1] found that the workpiece itself radiates most of the noise involved in cutting operations, especially where stainless steel and light metals are used. This noise may be reduced by loading the workpiece with damping plates, as shown in Figure 15.1. Only the free surfaces are used for damping the workpiece. The stiffness of the plates in combination with the loading used has to be adjusted to the stiffness of the workpiece. A typical damping plate construction is shown in Figure 15.2. Using this approach, a noise reduction from 108 dBA to 85 dBA was reported.

Tool chatter and also machining vibrations may often be reduced by the use of a lubricant or by modifying cutting parameters: cutting, speed, depth of cut, etc.

Gear noise may be controlled as discussed in Chapter 30; motor noise is discussed in Chapter 39.

Reference

1. Karlsson, S. O., "Noise Reduction When Milling Thin Walled Workpieces," Inter Noise 76 Proceedings, April 5-7, 1976, pp. 37-42.

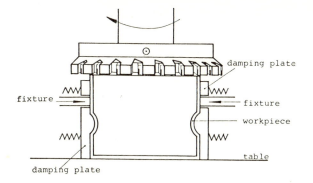

Figure 15.1. The damping principle with damping plates.[1]

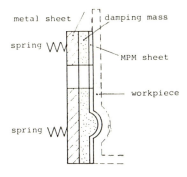

Figure 15.2. The construction of a damping plate.[1]

16 - DRILLS

Generally, very little noise is generated by drill-workpiece interaction. Where excessive noise is observed due to high speed operation, the use of a lubricant is recommended to reduce cutting forces. Occasionally resonant vibrations of the workpiece will generate high noise levels; this problem may be solved by applying a vibration damping treatment to the workpiece.

In most cases where excessive noise is observed for drilling operations, the primary cause is the gear drive system. Often, the problem is worn gears, and gear maintenance will solve the problem. Gear noise is discussed in Chapter 30.

For some operations, air is used for chip removal. This noise may be reduced by use of a thrust air silencer, as discussed in Chapter 28.

17 - GRINDING

A search of the literature on acoustics and noise control indicated that no analysis of grinding noise has been developed or published.

A general fundamental analysis of grinding noise is necessary in order to investigate potential modifications to the grinding operations as they may relate to noise reduction. In view of the apparent lack of published literature on the subject, a generalized analysis was made and a series of experiments were performed which are summarized here.

The energy associated with a grinding operation is transformed directly into three basic energy forms:

1. Molecular energy (material removal)

2. Heat energy

3. Vibrationan energy, which is transformed into:
   a. acoustic energy
   b. heat energy

Thus,

$$E_{total} = E_m + E_h + E_v$$

$$E_{total} = E_m + E_h + E_h´ + E_a$$

where: E_m = molecular energy
E_h = heat energy
E_v = vibrational energy
$E_h´$= vibrational heat energy
E_a = acoustic energy

The total energy required for a material removal operation may be estimated from motor power consumption. The energy required for material removal could be calculated from consideration of material strength properties. The vibrational energy induced into the system (both grinding wheel and part) is dependent upon the following parameters:

1. Material removal efficiency. Clearly the more efficient the material removal, the lower the resulting vibrations. For conventional grinding systems, however, the influence of efficiency is slight for relatively large variations in the grinding abrasive.

2. Amount of material removed.

3. Velocity of grinding surface.

VIBRATION CORRELATION

The near-field sound pressure levels due to vibrations induced by grinding may be calculated above the panel coincidence frequency from:

$$L_p = L_v - 20 \log f + 150$$

where: L_p = sound pressure level, dB
            L_v = vibration level, dB re 1.0 g
            f = frequency, Hz

The coincidence frequency for steel plates is:

$$f_c = 500 \div t$$

where: f_c = coincidence frequency
            t = thickness, inches

Below the coincidence frequency, the part sound radiation will be 3-4 dB/octave less than predicted by the equation. The equation is valid only for frequencies with wavelengths less than the part dimension.

Using this analysis, vibration levels of a part may be measured, and the investigator may determine if the primary noise source is radiation of sound from the part or some other machine element.

MATERIAL REMOVAL

An experiment was conducted on an O.D. grinder with a 3" wide, 0.4" thick strip of brake lining material to establish the influence of the amount of material removed on the sound level. An increased sound level with depth of grind was found, as shown in Figure 17.1.

SURFACE

An experiment was conducted on a 5" wide strip of brake lining material to establish a relationship between surface area of the vibrating strip being ground and sound level. The following sound levels were measured:

Surface Area (in²)	dBA
130	100
65	100
33	96

The lack of engineering feasibility for some grinding operations has been recognized by industry. In a recent study prepared for the U.S. Department of Labor, the acoustical consulting firm of Bolt, Beranek and Newman reported:

Examples of operations and equipment that might not be amenable to

sufficient noise control, either by engineering methods or administrative controls, include...certain grinding operations, such as
those involved in processing structural steel. ...

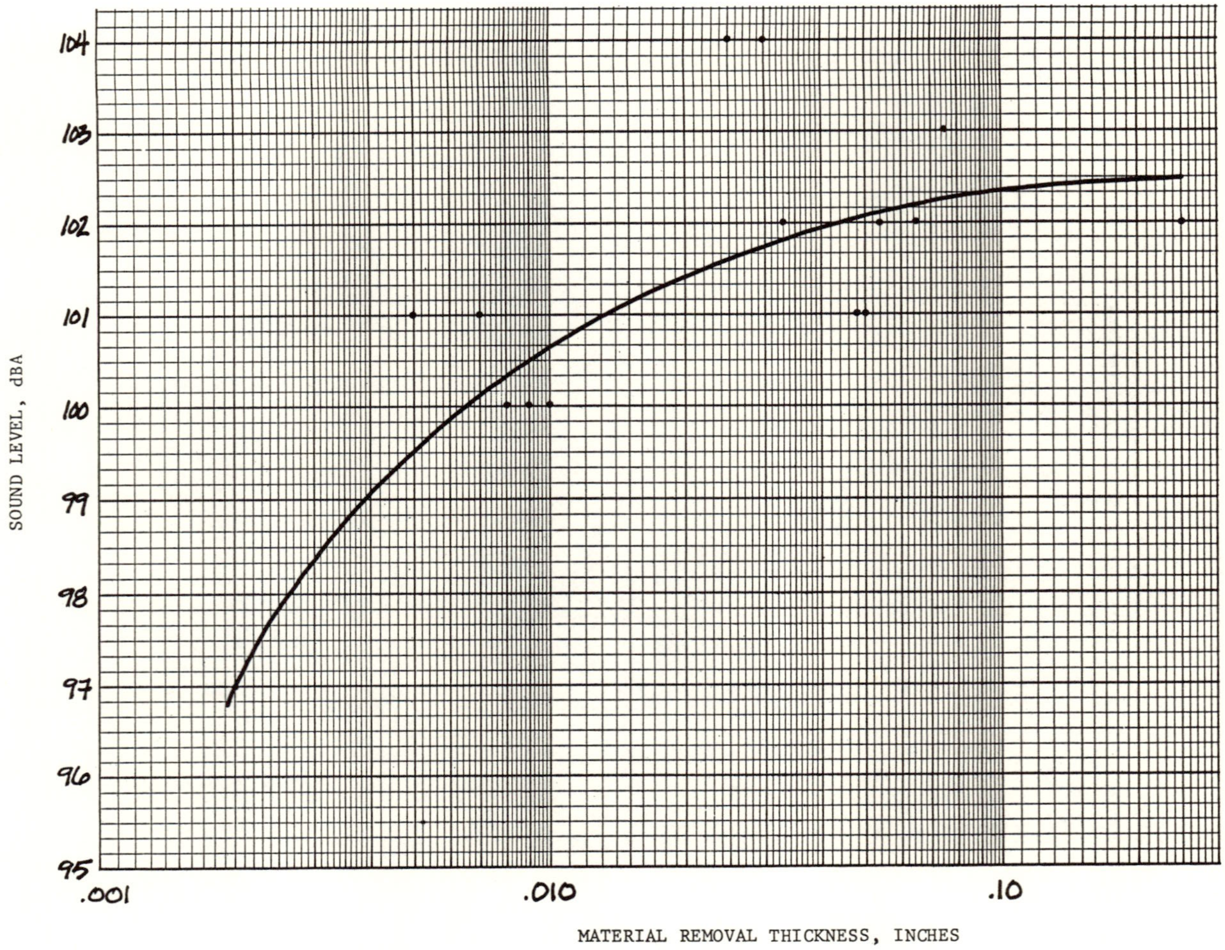

Figure 17.1. Experimental correlation between sound level and grinding material removal.

18 - AIR CARBON ARC GOUGING

Air carbon arc gouging (generally called arc airing or scarfing) is a manual operation which utilizes a carbon arc electrode and compressed air to remove welds and surface blemishes, or to prepare surfaces for welding. The electrode holder, which is hand held, is designed to retain the carbon arc rod and expel a stream of compressed air toward the arc. The carbon rod melts the metal and the air stream blows it away. In actuality, it is welding in reverse.

The noise associated with the scarfing operation is twofold. First, there is the noise due to an electrical spark discharge in air. The second noise source is the stream of compressed air used to remove molten metal.

Noise due to an electrical spark is a complex problem, due to the complexity of the source. An electrical spark generates acoustic N waves that have non-linear propagation properties. The non-linearity of the propagation of the N waves can have serious effects on common acoustical measurements of the spark. Adjusting the voltage used to create the spark will change the noise output. The spark gap dimension, d, will also vary the acoustical energy output. The efficiency of electrical to acoustical energy conversion increases at a rate proportional to $Log(d)$. In industrial applications, where large voltages are required and spark gap dimensions vary, control of these variables is not feasible as a means of noise control.

The air noise is secondary to the spark noise in a scarfing operation. The air noise generation is proportional to flow velocity to the eighth power. Thus, small reductions in velocity may result in significant noise reductions. Any such noise reduction, however, would not reduce the overall noise of the operation due to dominance of the noise resulting from the electrical spark.

The following efforts were made to identify potential solutions to the problem:

a. A literature search on welding.

b. Discussions with the largest manufacturer of arc air equipment in the United States.

c. Consulting studies with ten industrial users of the arc air process.

d. Discussions with members of the American Welding Society Project Committee on Noise (Richard K. Miller is a member of the committee).

These efforts failed to reveal any solutions to the arc air noise problem.

OSHA does not have any provisions for granting variances for non-feasible noise problems. Their procedure is that citations are not issued or are withdrawn when the lack of feasibility is recognized. Some states, however, which enforce OSHA do have such provisions for variances; in these states, variances for air carbon arc gouging have been granted. Various OSHA area offices have recognized the lack of feasibility by accepting compliance plan studies of arc air operations and allowing the use of a hearing conservation program as a permanent compliance measure. There are no decisions of the Occupational Safety and Health Review Commission (OSAHRC) which have recognized solutions to the problem.

Based on this analysis, it is concluded that there are no feasible engineering methods of noise control involving source modification for air carbon arc gouging within the existing state-of-the-art. It should be recognized, however, that in some plants, it may be possible to control employee noise exposure of the arc airing operations to some extent by workspace or administrative control techniques, such as:

a. Reverberation control
b. Localized barriers or shields
c. Reducing requirements for the process
d. Layout
e. Scheduling
f. Job rotation

19 - WELDING

With the exception of carbon arc operations (see Chapter 18) welding processes generally do not result in excessive employee noise exposure. This
is due to the fact that welding sound levels often do not exceed 90 dBA, and
also because welding operations are intermittent and exposure times are low.

The mechanism of noise generation has been deduced as being rarefraction,
ionization and/or chemical, and molecular decomposition of the air due to
the intense localized heat associated with welding. This theory has not
been confirmed, as no pbulished literature on welding noise could be found.

An extensive literature search was performed for this study to accumulate
resource information on state-of-the-art theories on welding noise. The
literature search included coverage of the eight references under the topics
of "welding", "noise", and "noise control", in the following references:

1. Science Abstracts, Part A.
2. Pollution Abstracts.
3. Noise and Vibration Bulletin.
4. American Industrial Hygiene Association Journal.
5. Applied Science and Technology Index.
6. Engineering Abstracts.
7. Environmental Health and Pollution Control.
8. Sound and Vibration Magazine Index.

Not a single paper on the subject could be found.

The only reference to welding noise was found in the *Handbook of Noise Control*[1], which listed the following sound levels associated with welding operations:

Welder, arc	80-89 dB
Welder, butt, electric	90-99 dB
Welder, gas, on steel	90-99 dB

During the course of our firm's studied we have measured the following:

Welder, stick	80-89 dBA
Welder, mig	85-102 dBA
Welder, carbon arc	102-118 dBA
Welder, carbon arc (air only)	92 dBA

The American Welding Society has established a Project Committee on Noise for
the purpose of:

> The establishment of the conditions under which noise measurements
> shall be made for the various welding, thermal cutting and allied
> processes, the collection of data and spectral characteristics of
> noise from a variety of welding, thermal cutting and allied pro-
> cesses and the evaluation of potential hazards to the hearing of

operating and nearby personnel, and the dessemination of findings as appropriate to regulatory agencies and others.

The committee has recently drafted a Noise Standard. It is not within the scope of this committee to consider work relating to the engineering control of noise, but rather simply to the measurement of noise and collection of data.

Reference

1. Harris, C. M., "Handbook of Noise Control," McGraw-Hill, 1974.

Vibrators are frequently for conveying materials and parts. The following approaches to noise control may be considered:

1. Install mufflers on pneumatic vibrators.

2. Substitute electric vibrators for noisier pneumatic units.

3. Use "cushioned stroke" vibrators.

4. Investigate alternate means of material handling.

5. Regulate vibrator use such that it is on only when required.

6. Operate vibrator at minimum required capacity.

7. Partially dampen system being vibrated.

8. Redesign system being vibrated to eliminate any large surface areas which would be prime sound radiators.

9. Enclose vibrator.

Noise control of vibratory bowl feeders is presented in Chapter 21.

CONVEYOR OPERATION

Guidelines for proper conveyor operation to insure minimum noise levels are presented in Table 20.1.

LIFT TRUCKS

An industry-wide (Industrial Truck Association) test procedure has been adopted which requires noise measurements to be made at the operator's ear plus 6, 12, and 18 feet from the side of the vehicle. These measurements are made at full speed, maximum load, and no load, plus during a "drive by". Many manufacturers specify sound levels in accordance with this standard.

Muffling of trucks may be accomplished by purchasing off-the-shelf mufflers and by shrouding the engine compartment. At present, fan noise is the major source of noise for LP gas vehicles, while high-speed DC motors are the major source of noise for electric vehicles. Power-steering pump noise also is a problem for the electric vehicles, but the noise of the electric vehicles is well below the OSHA requirements.

Reference

1. Wildsmith, C. G., "Better Operation of Conveyors and Elevators," *Food Manufacturer*, May, 1975, pp. 27-28.

TABLE 20.1

GUIDELINES FOR PROPER OPERATION OF CONVEYORS[1]

General

1. Are recommended lubrication procedures followed taking into account physical working conditions, intensity of operation and ambient temperature? And is the lubricant reaching the precise points where it is needed?

2. Is the equipment loaded in such a way as to avoid flooding with bulk materials and to reduce impact to a minimum, avoiding pulsations and surging?

3. Is the equipment always cleared before shutting it down? This avoids starting up under full load and resulting strain and wear.

4. Before starting up, is the equipment checked for obstructions or any undue build-up of materials – particularly those that tend to pack or harden – that could interfere with its operation?

5. Is the equipment run empty for a short period after starting up, to ensure fault-free operation and effective lubrication before being placed under full load?

6. Is the equipment run regularly – say once a week – during shut-down periods? This avoids risks of overloading through binding or seizing when the plant starts up again.

7. Are appropriate spares carried on site, to minimise downtime in the event of breakdown?

8. With outdoor installations in particular, is the supporting framework and casing, where fitted, kept painted against corrosion?

Drive gear

1. Are all the connected units correctly aligned? With V-belt drive, for instance, failure to ensure correct alignment of pulleys will result in rapid belt wear.

2. Is the drive gear inspected at least weekly for normal functioning without signs of overheating?

3. Is uniform tension applied to V-belts, with a slight 'bow' on the slack side of the belts when running under load?

4. Where oily conditions are encountered, are special V-belts fitted? Nor-

mally oil should not be allowed to come in contact with V-belts and belt dressing should never be employed on V-belts.

5. In the case of chain drives, is the slack side correctly tensioned? Over-tensioning will cause undue wearing of both chains and sprockets; too little tension may allow the chain to jump the sprocket teeth or the links to ride up the teeth, causing damage.

Chain conveyors

1. Is chain tension just sufficient to take up the slack and balanced where chains are used in parallel? Over-tensioning can create excessive wear.

2. Is the chain correctly aligned with the conveyor framework or casing and running centrally?

3. Is the chain's natural catenary interrupted at any point by rollers on the return strand? This can lead to high pin/bush wear.

4. Does the chain leave the driving and tail sprocket wheels cleanly without bunching between the wheels and chain support guides?

5. Are all rollers, where fitted, rotating freely and not sliding along their track supports?

6. Are chains used in sliding applications designed specifically for this type of duty?

7. Are sprockets of appropriate design, quality and not too small? The fewer the teeth the higher the 'chordal' action, producing increased pin/bush wear.

8. Are sprockets inspected regularly for signs of wear and to ensure correct meshing of the chain in the teeth? This affects gearing performances, jerky, irregular running causing rapid chain wear.

9. Are all attachment bolts securely tightened and damaged attachments replaced?

10. On overlapping slat or tray conveyors, does the slat profile allow sufficient clearance between the leading edge of the slat and the platform of the preceding slat? If not, or where other obstructions occur, severe stresses can be applied to the 'K' attachment of the chain – particularly where the return strands travel over rollers.

Belt conveyors

1. Are head and tail pulleys correctly aligned with each other and idlers also correctly aligned at right angles to the centre line of the belt, to ensure correct

tracking?

2. Has adjustment for correct tracking been carried out under loaded conditions, without undue reliance on 'knocking' or tilting of idlers, without unequal adjustment of take-up screws and without creating undue belt tension? All three of these methods of adjustment are incorrect and should be avoided.

3. Is applied tension just sufficient to prevent the belt slipping on the drive pulley under loaded conditions – and no more?

4. Are regular checks made to ensure that all idlers turn freely, any caked material being removed from idler and pulley surfaces?

5. Are belt ends joined by means of mechanical fasteners correctly cut (square with the centre line of the belt), checked regularly and fasteners replaced as necessary?

Snapping of worn fasteners results in uneven strain that can tear the belt lengthwise.

6. Are any stationary parts of the equipment (other than belt scrapers) or material wedged beneath feed chutes allowed to come in scraping contact with the surface or edges of the belt?

7. Are any cuts and abrasions in belt covers or edges repaired promptly as they are noticed?

Screw conveyors

1. Are flights checked regularly for signs of deformation which could in turn lead to serious damage to troughs and/or drive units?

2. Are intermediate hanger bearings checked at least monthly for signs of wear? Abnormally worn bearings will lead to the spiral shaft becoming bent or distorted and in turn cause the flights to rub on the bottom of the trough.

3. If a gudgeon and flight section are removed for servicing or replacement, is the spiral shaft checked to see that it is free to be turned by hand before power is applied to the motor?

Chain and bucket elevators

1. Is the chain correctly tensioned just to take up the slack, with both chains equally tensioned in the case of double strand elevators? Incorrect tensioning where buckets are of the dredging type will cause the bucket position to become uncontrolled, increasing wear on sprockets, chain, boot casing and at worst jamming the elevator and breaking the chain.

TABLE 20.1 (Continued)

2. Are all bucket bolts securely tightened and overlapping buckets capable of moving freely?

3. Are skidder bars, where fitted, clearing the inside of the elevator casing, replaced if worn or broken, and are the skidder bar bolts securely tightened?

4. Is the chain inspected at least monthly for correct meshing in the sprocket wheel teeth and any signs of abnormal wear?

5. Are buckets fouling the bottom of the boot, due to chain wear or 'stretch'? If so, clearance must be maintained by removing a few links or a bucket pitch of chain.

6. Are damaged buckets replaced as necessary? This applies particularly to those of the overlapping type, where deformation will impair operation of the elevator and may cause jamming.

7. Are checks made to ensure that material is not bridging in the feed or discharge chutes and that, with dredging-type elevators in particular, material is not allowed to build up in the boot above the normal running capacity of the elevator? Flooding of the boot causes excessive wear and tension in the chain, overloading of the drive gear and possible jamming of the elevator.

8. In outdoor installations in particular, is the casing kept painted against corrosion? Remember that with free-standing elevators in particular the casing serves a structural role and corrosion, if unchecked, can have serious consequences.

Belt and bucket elevators

1. Is belt tension just sufficient to prevent the belt slipping on the head pulley when running under load conditions? Belt slip results in excessive belt wear, reduced elevator capacity and eventual choking of the boot if the normal rate of feed is maintained. Also, sagging of the return side of the belt can result in 'flapping', causing interference of the buckets with the elevator casing.

2. Are the take-up screws equally adjusted? If not, the belt will be thrown out of line.

3. Are the belt and buckets inspected at least weekly to ensure that the belt is correctly tensioned, belt joint fasteners secure, all bolts kept tightened, worn washers replaced, correct clearances maintained and any faulty buckets replaced?

4. Are steps taken to guard against flooding of the boot? This can cause damage particularly when the belt is started, straining the belt and often resulting in slackening of belt tension.

5. Are the buckets running clear of the boot?

6. Are checks made to ensure that material is not bridging in the feed or discharge chutes?

7. Is the elevator casing kept well protected with paint? (See point 8 under previous heading).

21 - VIBRATORY BOWL FEEDERS

Vibratory bowl feeders are designed for a certain maximum design capacity, and are generally quiet within their design range. When this bowl speed is exceeded, excessive noise is generated due to system non-linearities. It is often observed that bowls are set to feed parts faster than required by the line flow. Maximum operating bowl speed should be determined for satisfactory operation and a limiter should be put on the potentiometer dial to prevent operation above this setting. The maximum operating point may be determined quiet easily by ear. If a limiter is not considered practical, the normal operational range should be marked on the potentiometer. In addition to noise level reduction, potentiometer control will result in both energy and machine wear savings.

Octave band sound pressure levels showing various modes of operation of vibratory bowls are shown in Figures 21.1, 21.2, and 21.3.

The sound levels of vibratory bowl feeders may easily be reduced to below 85 dBA by the installation of an acoustical enclosure. Two enclosure designs are shown in Figures 21.4 and 21.5. Noise control enclosures for vibratory bowl feeders are also available from the following manufacturers (cost generally ranges from $300-$800):

Ecology Controls, Inc.
223 Crescent Street
Waltham, MA 02154

Swanson-Erie Corporation
814 East 8th Street
Erie, PA 16512

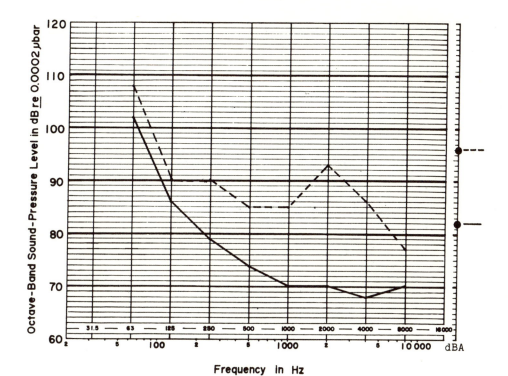

Figure 21.1. Vibratory bowl feeder without parts at normal potentiometer setting (————) and with potentiometer turned higher than design capacity (--------).

-83-

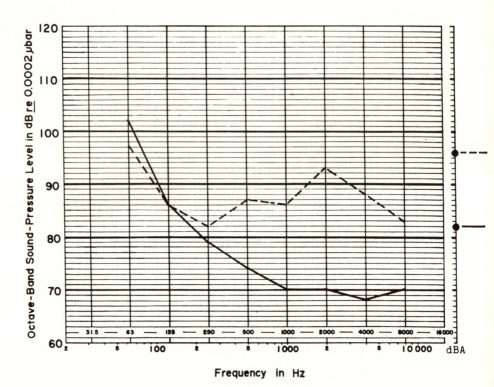

Figure 21.2. Vibratory bowl feeder empty (--------) and full (————) of parts.

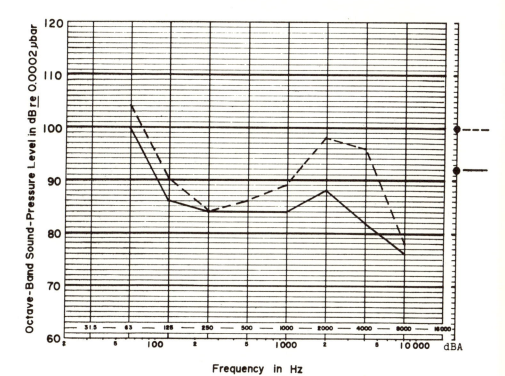

Figure 21.3. Vibratory bowl feeder during normal operation (--------) and with cardboard cover (————). Measurement location 6" above bowl.

1. Vibratory bowl feeder
2. Frame bolted to base for easy removal
3. 1.0 psf metal sides
4. Vibration damping material
5. 1" thick foam with 1 mil facing
6. Magnetic plastic gasket
7. Ferrous strip attached to cover
8. ¼" Plexiglas or clear acrylic

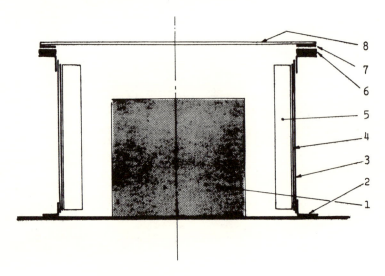

Note: Hinged and gasketed door may be
used in place of cover.

Figure 21.4. Acoustical enclosure with cover for vibratory bowl feeders.

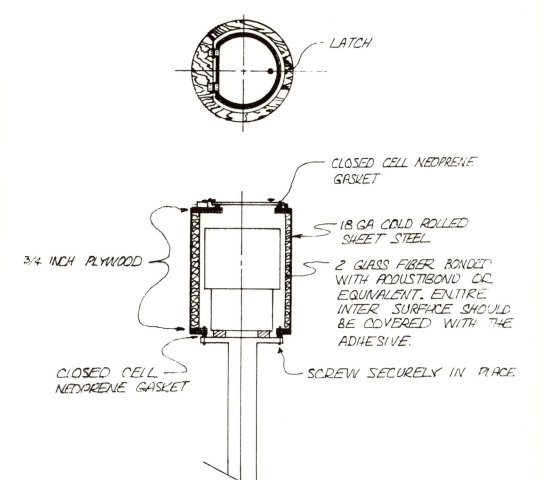

Figure 21.5. Acoustical enclosure for vibratory bowl feeder.

22 - HAMMERING

Impact noise is generated when metal parts are hammered to:

        a. Force parts into place.
        b. Finish weld surfaces.
        c. Shape parts.

This noise may potentially be reduced by:

1. Reducing the vibration response of the structure being impacted by means of vibration damping.

2. Reducing the impact forces.

3. Minimizing hammering requirements.

When parts are not of standard size or in a fixed location, vibration damping would generally not be practical. When operations are stationary, involve fixed dimension parts, and utilize production fixtures, vibration damping may be feasible.

The use of "soft" hammers, with plastic or rubber coated heads would serve to reduce impact forces. This approach has been tried in some industries. Often this substitution does not provide sufficient energy to perform the hammering function; however, it has worked in some cases. One such hammer is made of "compothane" which features a unicast head which can't fly off, mark, or spark. It is available from Compo-Cast, 2222 N. Olney Street, Indianapolis, IN 46218, and is said to outlast rubber, nylon, rawhide, or lead hammers by 10-20 times.

One plant which manufactures truck bodies has reduced the requirements for hammering by substituting a hydraulic frame press.[1] It is evident that this solution would not be applicable to every operation; however, it may be investigated.

In some cases, the necessity for hammering may be minimized by:

        a. The use of alternate assembly methods.

        b. Improvement of weld quality.

        c. Designing components to assemble better by means of better control of tolerances.

Reference

1. Kline, E.W., "Truck Factory Noise - A Better Way," Noisexpo 74 Proceedings, May, 1974, p. 231.

23 - SHOT AND ABRASIVE BLASTING

A recent report was pbulished by NIOSH which verified that, under current Occupational Safety and Health Administration standards, blast operations without noise protection using hand operated nozzles generally would be permitted less than two hours of noise exposure per day. If a survey indicates that the noise levels are excessive, the employer should implement a hearing conservation program consisting of engineering and administrative controls, use of personal protective equipment, and establishment of an audiometric testing program, as advised by the report. NIOSH said numerous scientific reports have shown that personal ear protectors, such as ear plugs or ear muffs, can provide 20 to 40 dB of protection if properly worn. To alleviate this noise hazard, the report suggested that operators wear muffs or ear plugs when wearing the helmets. The report (HEW Publication No. NIOSH 76-179) was prepared by Enviro-Management and Research, Inc., of Washington, D.C., under contract with NIOSH. Copies are available from NIOSH, Robert A. Taft Laboratories, 4676 Columbia Parkway, Cincinnati, Ohio 45226.

The major noise source where abrasive and shot blasting is performed is the compressed air and abrasive leaving the nozzle. The noise created by high velocity turbulent flow from an air jet is a function of air velocity and air nozzle diameter.

NOZZLE DESIGN

Until recently, there appeared to be no nozzle design modification which could be considered to provide feasible noise reduction. Differences in sound generation between 1/4" and 1/2" nozzles are slight. Nozzles of 1/8" diameter do not provide adequate abrasive flow for most applications. Pressure reductions from 100 to 80 psi do not result in any noise reduction; below 80 psi, blasting functions are seriously hampered.

Recently, Dr. John E. Sneckenberger of West Virginia University has undertaken considerable research into abrasive nozzle design. One design, an expanded barrel nozzle has been granted U.S. Patent No. 3,982,605, and noise reductions of 9 dBA are reported. The patent assignee, Pangborn Division, The Carborundum Company, is presently field testing the new nozzles. However, neither acoustical data from the field tests not any commercial products relating to the design are available at this time.

ABSORPTION IN BLAST ROOMS

It is evident that sound levels build up due to reverberation within blasting rooms due to hard wall surfaces. To determine the effect of the installation of sound absorptive material within a blast room approximately 20' x 10' x 8' in size, the ceiling was treated with a layer of 3½" glass fiber. Statistical sound measurements were made for an 8 hour work day at a location approximately five feet from the operator. The mean sound level was reduced from 107 dBA to 102.5 dBA, and the daily noise dose was reduced from 118% to 45%.

The problem is that the glass fiber must be protected from erosion by the blasting in the severe environment of the rooms. This is most difficult, if not impossible. Placing a solid rubber sheet in front of the glass fiber would completely negate its acoustical properties. One solution is to face the glass fiber with a screen and to replace the glass fiber frequently. Other possibilities would be to face the fiber with a perforated sheet metal or perforated rubber sheet (the holes should provide a minimum 15% open area).

BLAST HOODS

Typical blast hoods reduce noise levels at the employee's ear from 10-20 dBA. Thus, the proper use of ear plugs in conjuction with a blast hood would provide adequate protection of the operator in the absence of feasible methods of reducing the blasting noise.

Title 30, CFR, Subpart J, Section 11.120, requires that noise levels generated by the respirator, measured inside the hood at maximum air flow, not exceed 80 dBA.

A special blasting helmet, Model 2800AA has been developed with special sound attenuation features by:

> 3M Company
> St. Paul, MN 55101

24 - CUT-OFF SAWS

Noise from cut-off saws comes from two vibrating sources:

1. The saw blade.

2. The workpiece, which is usually largely unconstrained.

Since both the workpiece and blade are noise contributors, an acoustical enclosure is generally the best approach to noise control.

One enclosure design by Industrial Acoustics Company[1] is shown in Figure 24.1. Workpieces pass transversely through slots in the enclosure. Flaps of lead-loaded vinyl close off the opening and reduce to a small amount the unavoidable leakage area when a workpiece is present. The front, above saw bed height, is closed by two doors whose surface is mostly 1/4 inch clear plastic (polymethylmethacrylate). This plastic provides very good vision. The doors close with a gap the width of the control lever. Each door has a flap of lead-loaded vinyl about 3 inches wide to close the gap. The lever pushes aside the flaps only where it protrudes. Thus, the leakage toward the worker is greatly reduced. A 13 dBA noise reduction was achieved. The noise spectra at the worker position before and after the enclosure was installed is also shown in Figure 24.1.

For operations involving the sawing of thin-walled parts, it may be possible to design damping fixtures to be applied to the part, as discussed in Chapter 15.

Where the noise generated by the workpiece can be reduced, the blade must be treated also. This may be accomplished by installing a large stiffening collar on the blade, extending as near the base of the teeth as allowable by the stock thickness (the collar must cover 30% of the blade to be effective). A layer of vibration damping material, 1/8" thick, such as EAR C-1002, available from:

> Doug Biron Associates, Inc.
> P.O. Box 413
> Buford, GA 30518

should be installed between the collar and blade.

As an alternative to the above, the center portion of the blade not utilized for cutting may be treated with a constrained layer damping treatment. This treatment should cover as large an area as possible (30% is the minimum for effectiveness), and must not come into contact with the wood being cut. The treatment consists of a sandwich structure with a 10 mil DYAD viscoelastic layer, available from:

> The Soundcoat Co., Inc.
> 175 Pearl Street
> Brooklyn, NY 11201

and an outer layer of 18 GA steel. The construction should be bonded with an epoxy film.

A recent invention by Dr. Clayton H. Allen shows great promise in providing a technique for damping circular saws. The new damping means attaches to the saw table or base and employs low pressure air from a shop air supply or a small blower to force a pair of dampers against the sides of the saw in much the same way as disc brakes work. Noise levels peaking between 103 and 108 dBA have been reported to have been decreased to approximately 90 dBA when the damper was applied to the saw. The device is still in the prototype stage, and plans for commercial production are now being studied. Additional information on this design is presented in the literature from Brunswick Corporation, One Brunswick Plaza, Skokie, Illinois 60076.

Reference

1. Handley, J. M., "Noise - The Third Pollution, IAC Bulletin 6.0011.0, 1973.

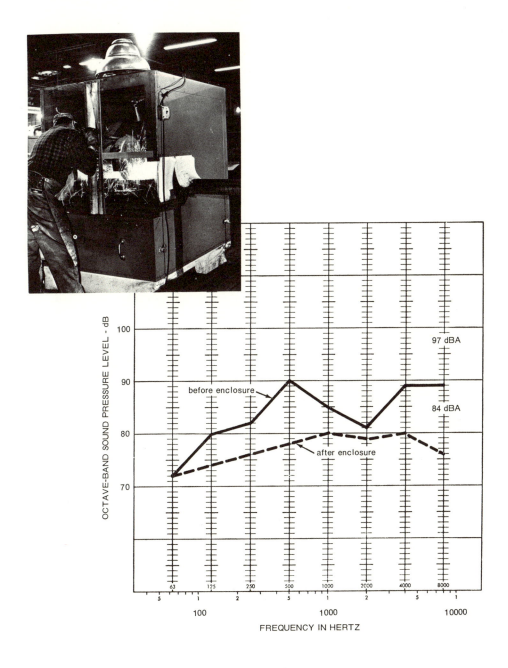

Figure 24.1. Metal cut-off saw: operator's exposure before and after enclosure of saw.[1]

-93-

Figure 24.2

   24" Diameter Cut-Off Saw Which Reduced Noise from a Reported
   120 - 130 dBA to 88 dBA.

Figure 24.3

   Acoustical Enclosure Constructed of Lead/Vinyl Curtain to Isolate
   Cut-Off Machine Noise from Adjacent Workers (Courtesy of Singer
   Partitions, Chicago, Ill.)

25 - BAND SAWS

Noise may be generated when cutting metal with a band saw in three ways:

        a.   Vibration of the saw blade.
        b.   Chatter between the part and saw table.
        c.   Vibration of the workpiece.

Generally the peak frequency of the noise is the tooth passage frequency, equal to the product of the blade speed (inches per second) and the number of teeth per inch. Thus, variation of blade speed provides one potential method of noise reduction; however, it is seldom possible to alter the speed sufficiently to achieve a significant noise reduction. A slight (1-2 dBA) noise reduction may be achieved by the installation of rubber facings covering the pulley wheels and/or blade guide wheels. Enclosure of the portion of the blade not involved in the actual cutting may provide additional noise reduction.

The saw damper invented by Dr. Allen (Chapter 24) is also applicable to band saws.

Noise produced by chatter between the workpiece and the saw table may be reduced by the placement of a wear resistant rubber material (Chapter 41) on the table. This phenomenon occurs most frequently in cutting large lightweight parts which bend due to lack of rigidity during cutting.

The most common source of noise in bandsaw operations is vibration of the workpiece. This noise may be reduced by the application of a damping plate to the part, as discussed in Chapter 15.

26 - SHEARS

The noise level of the shear operation is the result of four sources:

    1. The hold down impact on the stock.

    2. The impact of the blade on the metal.

    3. The stock "slap" and vibration on table after shear.

    4. Part drop impact.

There are two methods for reduction of the noise from the hold down impact on the stock. The hold downs may be covered with a wear resistant rubber material, or the control cylinder should be adjusted to decelerate the hold down descent.

Blade impact forces generally do not generate excessive noise on manually fed shears. Blades which cut at an angle reduce peak shear forces and are quieter.

Blade impact noise may exceed 100 dBA on high speed continuous feed shears. Since these shears do not require constant attention, noise exposure may most easily be controlled by isolating the operator. The noise may also be reduced by placing the blade at a slight angle or by neans of a machine enclosure or cover.

The noise generated by stock "slap" and vibration may be reduced by the shear table with a wear-resistant vibration damping material and by maintaining the hold downs in proper operating condition. Restraining rollers as shown in Figure 26.1 may also be used.

The noise producing impact forces are proportional to the kinetic energy, 1/2 mass (velocity)^2, which is proportional to the drop height. Therefore, reducing the drop height will result in a sound level decrease. Effective monitoring of drop heights at both the initial cut drop and stacking drop is necessary to control the sound levels of the operation.

The drop panel may also be lined with a wear-resistant rubber such as old conveyor belting.

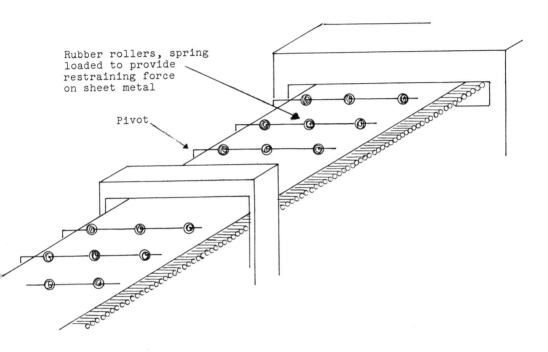

Rubber rollers, spring
loaded to provide
restraining force
on sheet metal

Pivot

Figure 26.1. Restraining rollers for McKay shear.

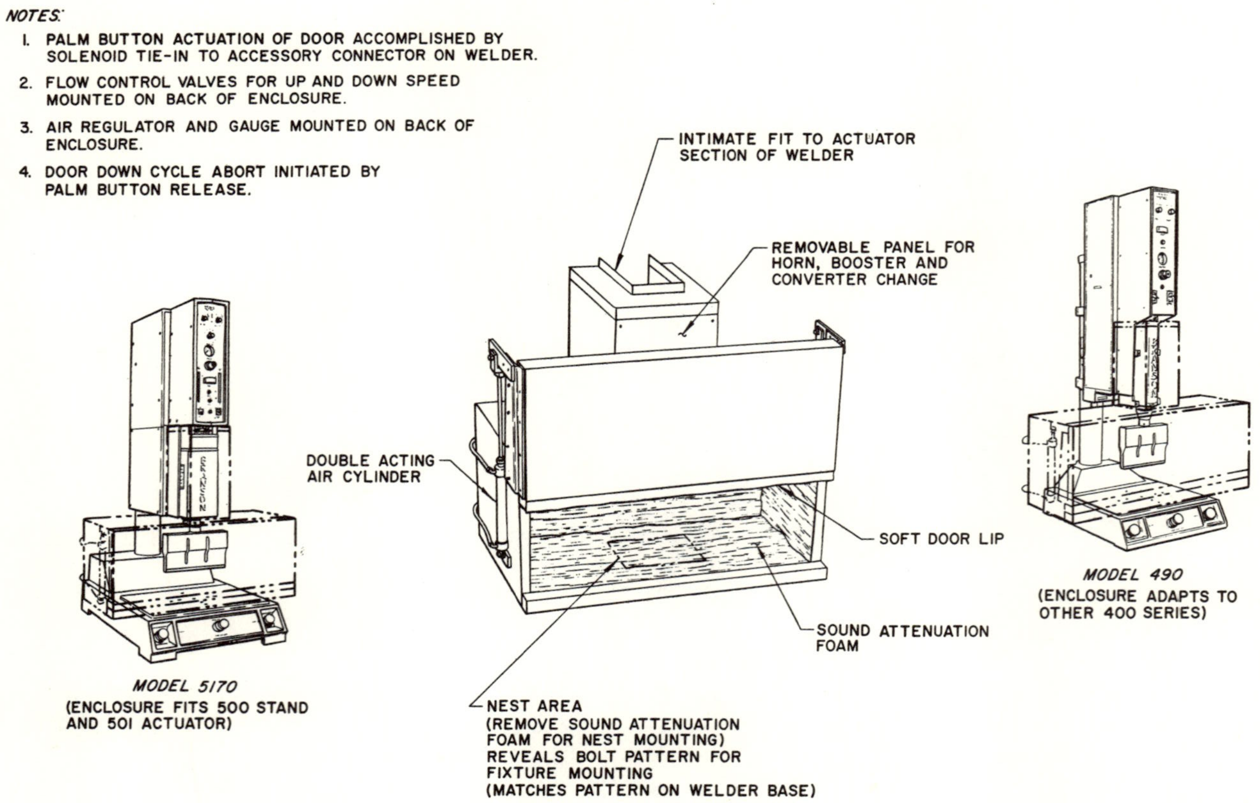

NOTES:

1. PALM BUTTON ACTUATION OF DOOR ACCOMPLISHED BY SOLENOID TIE-IN TO ACCESSORY CONNECTOR ON WELDER.

2. FLOW CONTROL VALVES FOR UP AND DOWN SPEED MOUNTED ON BACK OF ENCLOSURE.

3. AIR REGULATOR AND GAUGE MOUNTED ON BACK OF ENCLOSURE.

4. DOOR DOWN CYCLE ABORT INITIATED BY PALM BUTTON RELEASE.

INTIMATE FIT TO ACTUATOR SECTION OF WELDER

REMOVABLE PANEL FOR HORN, BOOSTER AND CONVERTER CHANGE

SOFT DOOR LIP

SOUND ATTENUATION FOAM

NEST AREA
(REMOVE SOUND ATTENUATION FOAM FOR NEST MOUNTING)
REVEALS BOLT PATTERN FOR FIXTURE MOUNTING
(MATCHES PATTERN ON WELDER BASE)

DOUBLE ACTING AIR CYLINDER

MODEL 490
(ENCLOSURE ADAPTS TO OTHER 400 SERIES)

MODEL 5170
(ENCLOSURE FITS 500 STAND AND 501 ACTUATOR)

Figure 27.1

Acoustical Enclosure for Ultrasonic Welder, Automatic, Palm Button Activated (Courtesy of Branson Sonic Power Company).

27 - ULTRASONIC WELDERS

The noise from ultrasonic welders may range from 80 dBA to 110 dBA. Due to the intermittent nature of this noise, welders seldom create an OSHA exposure problem. However, the very high frequency audible squeel which is emitted may require abatement due to annoyance considerations.

The first approach to sound reduction is to insure that everything possible has been done to eliminate the offending sound via control settings, fixturing or nexting. This may include everything from varying pressures, times and amplitudes to part clamps or nodally mounted horn clamps.

Where this approach does not provide adequate reduction, an acoustical enclosure may be employed. One design is shown in Figure 27.1. The unit is available from Branson Sonic Power Company, Danbury, CT. Noise reductions from 110 - 120 dBA to 85 - 90 dBA have been reported, and costs range from $1,000 to $1,500.

It is often asked whether ultrasonic sounds are more damaging to humans than audible sounds. In assessing available research on ultrisonic sound, the U.S. Environmental Protection Agency concluded that:

"Exposure to high levels of ultrasound (above 105 dB SPL) may have some effects on man; however, it is important to recognize that a hazard also arises from exposure to the high levels of components in the audible range that often accompany ultrasonic waves. At levels below 105 dB SPL there does not appear to be significant danger." [2]

Reference

1. U.S. Environmental Protection Agency "Public Health and Welfare Criteria for Noise" 550/9-73-002, July 27, 1973.

28 - THRUST AIR NOISE

In applications which require a jet of air to do useful work, effective thrust must be maintained to transfer force to the object. One such application is the ejection of parts from press dies, where open air jets are typically used. This noise may be reduced by the following methods:

1. Air ejector mufflers
2. Directional control of air flow
3. Regulation of air
4. Barriers
5. Mechanical ejectors

The thrust from an exit airstream is given by the equation:

$$T = \frac{W \cdot u}{g}$$

where: W = weight flow rate (lb/sec)
u = airstream velocity (ft/sec)
g = acceleration due to gravity = 32.2 ft/sec^2

The acoustic power (AP) in watts of an exit airstream is given by the relationship:

$$AP \propto \frac{Wu^7}{2gc^5}$$

where: c = velocity of sound (ft/sec)

With a seventh power dependency on air velocity associated with noise generation, but a direct relationship between thrust and velocity, an effective way to reduce high noise levels from nozzles is to obtain a nozzle which would exhibit somewhat lower exit velocity and reduce turbulence while maintaining high effective thrust. Several commercially available silencers have been developed to serve this function.

A recent study showed three commercially available silencers capable of producing thrusts of 32 ounces with sound levels from 5 to 10 dBA less than those associated with a free air jet of identical thrust.[1] These silencers are (in order of quietest first):

Model: E-2-6 Ejector (1/4")
Available from: C. W. Morris Company
36740 Commerce Street
Livonia, MI 48150

```
         Model:  Type AE (1/4")
Available from:  Allied Witan Company
                 12500 Bellaire Road
                 Cleveland, OH  44135

         Model:  309 (3/8")
Available from:  Sunnex
                 87 Crescent Road
                 Needham, MA  02194
```

Where there is not room for a muffler, the use of a 1/4" AE type Collimator insert, available from Allied Witan (address above) may be considered.

Air turbulence and resulting noise is created when high velocity air passes over irregular surfaces such as the edge of a die. Air jets should be directed toward die surfaces where minimum irregularities are encountered. It should be pointed out that the use of ejector silencers will also reduce noise of this type, since air flow is more directional with the silencers.

Air jets are required to eject parts only during a portion of a press cycle; however, continuous air flow is used on many presses. Air flow should be regulated to operate only during the part ejection portion of the cycle. In addition to noise reduction, a potential annual cost savings of up to $820 per press may be realized (assuming 1/4" air jet, 100 psig, continuous 8-hour per day operation, 10% ejection cycle time, and an air cost of 0.07 per 1000 cubic feet).

It should also be recognized that localized shields which block the line-of-sight path between a high frequency noise source such as an air jet and an operator may reduce sound levels by approximately 5 dBA. Such shields may be constructed of transparent plastic materials.

As an alternative to part ejection by air, mechanical ejectors may be installed. This approach would require significant engineering design, and may create maintenance problems, however, may achieve cost savings on a long term basis.

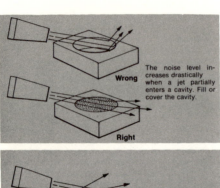

The noise level increases drastically when a jet partially enters a cavity. Fill or cover the cavity.

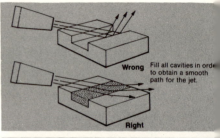

Fill all cavities in order to obtain a smooth path for the jet.

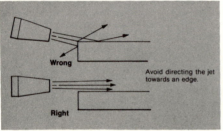

Avoid directing the jet towards an edge.

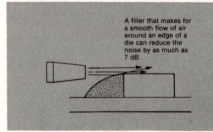

A filler that makes for a smooth flow of air around an edge of a die can reduce the noise by as much as 7 dB.

Figure 28.1

Reducing Air Turbulance Noise by Eliminating Obstacles in Path of Jet (Courtesy of Sunnex, Needham, Mass.)

29 - PART CLEANING AND DRYING

Sound levels of 95-100 dBA are typical of tumbling barrels used to deburr small parts. The majority of the noise is due to part-part impacts. This problem may be solved by simply installing a lid or cover on the mouth of the tumbler. Typically a lid may be fabricated of 1/2" to 1-1/2" plywood carefully fitted and tightly gasketed. Plastic covers resembling kitchen bowl covers have also been used successfully. One plant incurred a serious problem in covering their tumblers: the stainless steel parts which were being deburred became tarnished. The problem was deduced as being due to moisture which became trapped in the closed system, and the problem was remedied by adding corn cob to the tumbler.

Often, abrasive stone is added to the tumbler. The result is a very significant decrease in part-to-part noise, however, the barrel itself will have an increased tendancy to radiate noise due to vibrations induced by the part/ stone fill hitting the steel walls. This noise problem may be solved by either lining the interior with a wear resistant rubber material, applying a vibration damping compound to the barrel, or by lagging the barrel.

The combination of the two treatments listed above have reduced tumbler noise from 98 dBA to 78 dBA.

Vibratory type tumblers present another potential noise problem. The part-to-part component of the noise may be reduced by using a high abrasive stone-to-part ratio. The mechanical condition. Unbalances, bent shafts, loose parts, etc. can cause a rise in sound levels of 5 dBA or more. Unfortunately, in such a vibratory system, this mechanical noise recurs shortly in such a vibratory system, this mechanical noise recurs shortly after maintenance, and the frequency of maintenance may become greater than is justifiable for noise control. The only other solution is to use an acoustical enclosure or another type of vibrator.

Part Drying

In some metal fabrication operations, water is used to wash or cool parts. Generally, the water is removed by either compressed air jets or by an air curtain (a pipe with a series of air holes). Noise levels above 100 dBA at 3' are typical. The mechanism of drying is generally blowing the water from the part rather than evaporation.

Where parts are irregularly shaped, the following approached may be considered:

1. The high pressure air system may be replaced by a high volume, low pressure system, such as supplied by a small blower.

2. A Transvector air nozzle or air curtain may be used, such as available from:

    Vortec Corporation
    4511 Reading Road
    Cincinnati, OH 45229

3. A silencer designed to provide concentrated air flow may be used (see Chapter 40).

4. The air or part may be heated to increase the evaporation rate.

5. The operation may be enclosed.

For regularly shaped parts, the above approaches may also be considered; however, the best solution is to replace the air drying system with mechanical wipers. In addition to providing virtually silent operation, this approach may save thousands of dollars in compressed air costs annually (see Chapter 28).

30 - GEARS

Geared systems may be extremely noisy. Gears consist of assemblies of toothed wheels used for the purpose of torque conversion, speed change, or power distribution. The main sources of noise in geared systems are:

       1. Impact caused by tooth contacts.

       2. Mechanical imbalance of the gear assembly.

       3. Friction due to the contact motion of the tooth.

       4. Variation of radial forces.

       5. Air and oil pocketing.

Some of the principles used for reducing noise in gear systems are:

       a. Selection of a suitable type of gear (for instance, a helical gear is quieter than a spur gear, and a worm gear is still quieter, but is restricted to low speeds).

       b. Accuracy of manufacturing (high accuracy in all gear parameters results in quieter gear systems).

       c. Detuning (when the operational frequency of the gear assembly coincides with the natural frequency of the structural members, resonance takes place, amplifying the noise; to avoid resonance, the structural members are detuned to other frequencies by either stiffening or mass loading).

       d. Damping (introduced by using gear material of high internal damping).

       e. Vibration isolation.

       f. Enclosing the gear assembly (with particular attention given to cooling and heat transfer requirements).

Table 30.1 provides the noise reductions that are possible by appropriate adjustment of design parameters.[2]

The use of rawhide, nylon, and laminated phenolic gears is often proposed as a method of noise reduction. The Lewis formula may be applied to assess gear strength requirements:

$$X = \frac{D^2 FR}{50\ HR}$$

Where: D = outside diameter (inches)
        F = race width (inches)
        R = rotations per minute
        H = horsepower transmitted
        N = number of teeth

Non-metallic gears should not be used for applications where the above relationship yields solutions of less than 1.0.

TABLE 30.1

AVAILABLE NOISE REDUCTIONS FOR GEARED SYSTEMS[2]

Design Parameter	Noise Reduction in dB	Remarks
Profile Error	0-5 5-10	Normal Manufacturing Ultra Precision Gears
Profile Roughness	3-7	Full Range of Standard Manu- facturing Techniques
Tooth Spacing Error	3-5	
Tooth Alignment Error	0-8	
Speed	$\approx 20 \; log \; (\frac{V}{V_o})$	Basic Data V = Speed
Load	$\approx 20 \; log \; (\frac{V}{L_o})$	Basic Data, High Loads and Speeds L = Load
Power	$\approx 20 \; log \; (\frac{LV}{L_o V_o})$	Basic Data
Pitch	Not Known	Finer, Quieter
Contact Ratio	0-7	Largest Best, But if Small Contact Ratios are Necessary, Use 2.0
Angle of Approach and Recess	Not Known	Approach Forces Higher... Smaller Approach Angle Quieter
Pressure Angle	Not Known	Lower Pressure Angle, Quieter
Helix Angle	2-4	For Changes from Spur to Helix

TABLE 30.1 (Continued)

Design Parameter	Noise Reduction in dB	Remarks
Gear Tooth Backlash	0–14 3–5	If Excessive Backlash If Too Little Backlash
Air Ejection Effects	6–10	5000 fpm or More
Tooth Phasing	Not Known	Not Practical
Planetary System Phasing	5–11	Practical
Gear Housing	6–10	If Resonant
Gear Damping	0–5	If Resonant or Needs Isolation
Bearing	0–4	Adds Damping, Some Types May Stiffen Structure
Bearing Installation	0–2	Can Increase Life and Eliminate Some Frequencies
Lubrication	0–2	Filled Gearbox Quietest, but Can Cause Other Problems

31 - SHEET STACKING AND UNSTACKING

The fabrication of many parts begins with a metal sheet as the raw material. Initial production operations involve unstacking and stacking of the sheets. Noise may be generated due to the following:

1. Unstacking: Air jet used for sheet separation.

2. Stacking:

   a. Impact of sheet on rigid stop or metal inertia block.

   b. Impact of sheet on stack.

The following approaches to noise control may be considered:

UNSTACKING

An open air jet is often used to separate sheets as they are unstacked, generating noise levels from 95 to 105 dBA. The use of a thrust muffler (Chapter 40) would not be effective since the dominant noise generating mechanism is turbulence generated as the air passes the sheet (dipole noise) rather than noise from the free air jet (quadrupole noise). A silencer would not significantly alter the air-sheet interaction.

A very attractive solution to this problem both from a noise control and energy conservation point of view, is to replace the air separator with a magnetic fanner, such as aviable from:

> Dura Magnetics, Inc.
> Pyle Drive
> Sylvania, OH 43560
>
> Magnetool
> 1051 Naughton
> Troy, MI 48084
>
> Industrial Magnetics, Inc.
> 1462 E. Big Beaver Road
> Troy, MI 48084
>
> Permag
> 1919 Hills Avenue, NW
> Atlanta, GA 30318
>
> Hitachi Magnetics Corp.
> Edmore, MI 48829
>
> Homer Magnetics
> 915 Shawnee Road
> P.O. Box 386
> Lima, OH 45802

The only other approach to noise reduction would be the use of an enclosure or barrier.

SHEET IMPACT WITH STOPS

As the sheet leaves a conveyor line, it will be decellerated by a positioning stop prior to dropping onto a stack. Generally, the stop is metal and the impact induces noise producing vibrations into the sheet. This noise may be reduced by installing a wear retardant rubber facing on the face of the stop (see Chapter 41). The stop may also be spring loaded to provide additional cushioning.

SHEET DROP ONTO STACK

Excessive noise may be generated as the sheet free falls onto a stack. The noise generated is directly proportional to the drop height (potential energy = weight x height). Thus, one method of noise control is to minimize this distance.

Another approach is to install a bin with four solid sides tightly around the stack. As the sheet enters the bin, air will be trapped between it and the stack, allowing the sheet to float quietly into position.

Impact noise generated when parts are dropped into tote bins may exceed 115 dBA. Although this is a short duration noise, it may add considerably to the background ambient in work areas. Many plants have hundreds of tote bins; thus any noise control solution must be applicable for implementation for a large number of units.

The following solutions may be considered:

1. The interior floor may be lined with a wear retardant material to cushion part drops (see Chapter 41 for materials list). Wood and sand have also been used in some plants for this purpose.

2. Replace solid metal bins with heavy duty wire mesh bins. This approach may have several non-acoustical benefits:

   a. They are lighter weight.

   b. Space savings (the bins may be stacked).

   c. Visibility through empty bins.

3. The exterior of the bins may be treated with a vibration damping such as EAR-1003 (see Figure 32.1), available from:

   > Doug Biron Associates, Inc.
   > P.O. Box 413
   > Buford, GA   30518

   or a spray-on or trowel-on damping compound.

4. Where hot parts are dumped into bins, items 1 and 3 above are not applicable. An alternate design is to install a wire mesh false floor in the bin, as shown in Figure 32.1, to isolate the dropping parts from the bin panels.

(a)

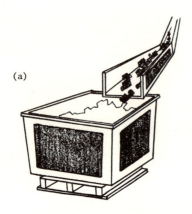

(b)

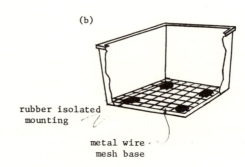

rubber isolated
mounting

metal wire
mesh base

Figure 32.1. Noise control solutions for tote bin impact noise: (a) exterior
vibration damping; (b) isolated wire mesh floor.

33 - IMPACT RIVETING

Impact riveting operations are responsible for some of the highest noise levels found ing American industry; up to 120 dBA. Impact riveting involves the use of a pneumatic hammering tool which forms the rivet head by means of repeated impacts.

Three conceptual approaches to noise abatement of the impact riveting operations may be considered:

1. Substitution of fastening process.

2. Isolation by relocation or rescheduling.

3. Vibration damping of workpiece.

SUBSTITUTION

Noise reduction could be achieved from substitution of impact riveting processes with another fastening process, such as hot or cold squeeze riveting, Huck bolts, bolts, or welding.

A new fastening system has been developed by Fraser Automation Co., 31125 Fraser Drive, Fraser, Michigan, to replace some impact riveting operations. The system uses an orbital and planetary action which can form flat, crown, and conical heads. It is virtually silent.

ISOLATION

Isolating riveting operations from employees not physically involved in the process may be considered by:

a. Physical location of the riveting fabrication areas within a separate building or work area.

b. Performing riveting operations during nighttime shifts when other workers are not present.

DAMPING

A significant noise reduction may often be achieved by positioning a vibration damping material in contact with the work piece being riveted. As shown in Figure 33.1, partial coverage of a panel provides damping efficiencies far in excess of the percent of area covered. It is most important that each separate panel or structure attached to the part being riveted be damped in order to achieve significant noise reduction. A limitation in noise reduction is the fact that vibrations are attributed to both forced impacts to the panels as well as resonant panel response. Vibration damping is only effective in reducing resonant vibrations, and the impact mechanism will inherently result in a substantial vibrational energy which cannot be reduced by means of vibration damping in any manner.

Impact Riveting

LIMITATIONS

It should be pointed out that impact riveting operations cannot feasibly be replaced or abated for some operations because of several reasons, including:

    a. The superior strength of hot impacted rivets.

    b. The mobility of impact riveting equipment as opposed to the very large size of squeeze riveting units.

    c. The high costs involved.

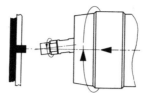

Pressure and orbital motion are both used to form accurately controlled rivet or fastener heads. The tool moves in an orbital path around the centerpoint of the part, creating a moving line of pressure.

Figure 33.1

New Fastening System

Planetary action means small, delicate parts can now be headed without undue galling or distortion. A series of rosettes is generated about the centerpoint of the rivet or stud. The head-forming tool is interchangeable with other heads for the machine.

Figure 33.2

New Fastening System

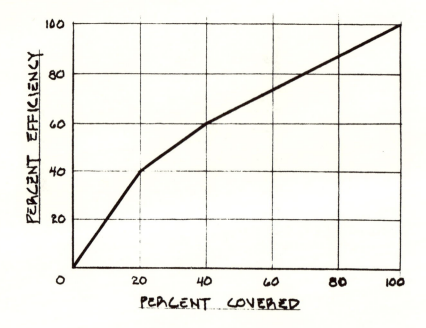

Figure 33.1. Effectiveness of a vibration damping treatment as a function of the area covered.

34 - CRANES

The operators of open overhead cranes are often exposed to excessive sound
levels by virtue of exposure to work areas over which the crane passes.
Where crane operators are found to be exposed to excessive noise, it is generally the case that floor level employees also encounter noise exposure.
In many cases, a plant-wide noise porgram will accomplish adequate noise reduction, and the exposure of the crane operators will be sufficiently reduced. Where crane operators remain overexposed to noise after plant machinery treatment, the crane cab may be modified.

Crane operators may be isolated from exterior noise by installing safety
glass or clear plastic windows in the cab. The windows should be operable or
angled and extend beyond the cab for visibility. Visibility should never be
sacrificed for noise control. With the use of windows, the cab should be
ventilated and probably air conditioned.

Additional noise reduction may be achieved by installing a sound absorptive
material on the interior ceiling and walls of the cab. A carpet may also be
used. These measures would provide thermal as well as noise control benefits.

Some cranes are found to generate noise themselves. In most cases, this is
due to the gear system, which may be treated as discussed in Chapter 30.

Sirens and whistles are utilized as warning signals throughout the mill. The sound levels of these sources must be high to insure detection and employee safety.

In many cases, the sound levels of a siren or whistle may be reduced while maintaining an adequate warning level if the following concepts were implemented:

1. The siren may be a warble, or intermittent tone, similar to British police sirens. The detection limit of an oscillating tone is greater than for a steady sound, and the overall sound level could be slightly reduced and the exposure time lowered.

2. The peak frequency may be reduced from 500 Hz to 250 Hz, since the ear is equally sensitive to 250 Hz as 500 Hz; however, the 250 Hz level of equal loudness would be 6 dB lower on the A-weighted scale.

3. The single siren may be replaced by two more directional sirens.

While the ear normally is most sensitive to sounds at 4 kHz, warning signals at that frequency are not recommended, since that range is also the characteristic hearing loss range due to noise exposure. Thus, workers with a PTS (permanent threshold shift) or TTS (temporary threshold shift) would not detect the high frequency siren noise as well as they would a low frequency source.

A final point is that the siren level can be reduced as the background noise level is decreased, since the detection level is proportional to the background noise due to masking effects.

References

1. Fletcher, H., and Munson, W. A., *Journal of the Acoustical Society of America*, Vol. 5, No. 82, 1933.

2. Robinson, D. W., and Dadson, R. S., *British Journal of Applied Physics*, Vol. 7, No. 166, 1956.

36 - MAN COOLER FANS

The movement of high velocity air across the body is the most effective method of reducing employee heat stress in the working environment. Creating a high air flow by means of fans is very critical in maintaining employee health and comfort. In an inventory of over 100 man cooler fans, sound levels ranging from 85 to 103 dBA were measured at 6', with the sound levels of 20% of the units exceeding 90 dBA.

Sound levels of the fand are influenced by the following variables:

- Speed or rpm of the motor
- Velocity output of the fan
- Number of blades
- Blade shape and pitch
- Horsepower
- Diameter

The following fan laws relate to noise generation:

$$CFM_a = CFM_b \times \left(\frac{size_a}{size_b}\right)^3 \times \left(\frac{RPM_a}{RPM_b}\right)^1$$

$$HP_a = HP_b \times \left(\frac{size_a}{size_b}\right)^5 \times \left(\frac{RPM_a}{RPM_b}\right)^3$$

$$SPL_a - SPL_b = 70 \log_{10}\left(\frac{size_a}{size_b}\right) + 50 \log_{10}\left(\frac{RPM_a}{RPM_b}\right)$$

Blade shape is an important factor in fan noise generation; a 10 dBA reduction may be achieved by blade shape modification.

Where excessive sound levels are observed from man coolers, the following approaches may be taken:

1. Replace man coolers with new quiet models. Approximate cost per unit is $300-$400.

2. Install new quiet blades on the man coolers, such as Robertson replacement blaced, available for approximately $60 for a 24", 4-blade fan and $110 for a 36", 4-blade fan.

If new fans are purchased, a specification similar to the following should be used:

> The vendor shall guarantee that the sound level of the fan does not exceed 85 dBA, measured at 6 feet.

A specification requiring that the fan "comply with OSHA" may not provide adequate results, as the manufacturer may measure his fan noise at a distance considerably greater than 6 feet.

References

1. Jorgensen, ed., *Fan Engineering*, 7th edition, Buffalo Forge Company, Buffalo, NY, 1970.

2. Lord, H. W., Evensen, H. A., Canada, I. C., and Palmer, M., "Effects of Modifications of Man Cooler Fans on Thermal and Noise Environments," proceedings of NOISEXPO, April 30 - May 2, 1975, pp. 65-67.

37 - VENTILATION FANS

The sound levels due to the roof ventilation fans often exceed 85 dBA, and occasionally exceed 90 dBA. Silencing of these fans is frequently required to achieve satisfactory ambient levels in a plant. Three approaches may be considered:

1. *Fan Silencers*. The installation of silencers is considered the most effective abatement method. The "straight-through" type should be used so as to minimize air flow impedence. Commercially available systems, such as from Industrial Acoustics Company, generally cost about $2000 per unit. Installation costs may be twice the unit cost.

2. *Replacement Blades*. There are commercially available replacement blades of a multitude of configurations for noise reduction purposes. Robertson Air Systems offers replacement blades that are compatible with most existing horsepowers, rpm's and sizes.

3. *Hanging Barrier*. Suspending a barrier directly below the fan, obstructing the line-of-sight to the workfloor may be considered. If a sound absorptive barrier of twice the fan diameter were hung 1½ times the diameter below the unit, a noise reduction of 1-4 dBA may be expected. A more restrictive barrier may provide noise reductions up to 10 dBA; however, such units are generally impractical, because a 10 to 15 percent impedance of the fan's air flow would result.

When purchasing new fan units, the engineer should always include noise level specifications.

38 - REVERBERATION

Whenever machinery is operated within enclosed spaces, sound levels will be increased to some extent due to reverberation. When this reverberant sound level increase becomes significant, it is appropriate to install sound absorptive materials on the ceiling above offending machinery. A simplified procedure can be used to estimate the increase in noise due to the reflected component, as shown in Figure 38.1.

Ceiling treatment is also required wherever acoustical barriers are employed to prevent sound from being reflected off the ceiling and over the barrier.

The most convenient method of employing sound absorption is the installation of acoustical baffles. The following is a list of manufacturers of commercially available baffle systems:

> Owens Corning
> Fiberglas Tower
> Toledo, OH 43659
>
> U.S. Gypsum
> 101 Wouth Wacker Drive
> Chicago, IL 60606
>
> Industrial Noise Control, Inc.
> 785 Industrial Drive
> Elmhurst, IL 60126
>
> Armstrong Cork Corporation
> Lancaster, PA 17604
>
> Eckel Industries, Inc.
> 155 Fawcett Street
> Cambridge, MA 02138

Where fire sprinkler systems or air and ventilation requirements preclude the use of baffles, 1" thick conventional acoustical tile may be installed directly on the ceiling.

Where large surface areas are to be treated, a spray-on treatment may be most economical. These materials are available from:

> National Cellulose Corporation
> 12315 Robin Boulevard
> Houston, TX 77045
>
> Sprayon Research Corporation
> 5701 Bayview Drive
> Ft. Lauderdale, FL 33308

Attention should also be given to machine location within a room. These relationships are shown in Figure 38.2.

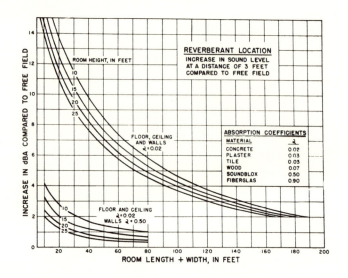

Figure 38.1. Simplified procedure for esitmating increase in noise due to reverberation.

Absorption Coefficient	Change Location	Increase in SPL
0.02	A to B	1.5
0.02	A to C	3.0
0.50	A to B	3.0
0.50	A to C	6.0

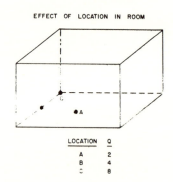

EFFECT OF LOCATION IN ROOM

LOCATION	Q
A	2
B	4
C	8

Figure 38.2. Effect on location in room.

With approximately two million motors sold a year, motor noise is of great concern. The two primary types of motor enclosures are the dripproof (DRPR) and the totally enclosed, fan cooled (TEFC) motors. The DRPR cools itself by pulling in outside air and circulating it around the electrical conductors, while the TEFC motor uses a fan to accomplish cooling.

Table 39.1 presents a comparison of National Electrical Manufacturers Association (NEMA) and International Electrotechnical Commission (IES) motor noise standards with those of a leading motor manufacturer. It is evident that high speed (3500 rpm and above) and TEFC motors are noisier than their slower speed and DRPR counterparts.

Solutions to motor noise problems include muting or muffling the inlet and outlet air and isolating or dampening the vibrations. Motors with horsepower ratings of 75 or more often require abatement. Motors below this rating do not normally generate noise levels that are above 90 dBA; however, these smaller motors may contribute to other equipment noise, and the additive levels may exceed 90 dBA. The following firms manufacture motor silencers:

> The Spencer Turbine Company
> 600 Day Hill Road
> Windsor, CT 06095

> Pulsco Division
> American Air Filter Company
> 215 Central Avenue
> Louisville, KY 40201

Several motor manufacturers have developed new lines of quiet motors with higher horsepower ratings. This has been accomplished through quiet fan designs and minor internal modifications. To insure minimum motor noise levels, specifications that sound pressure levels are not to exceed 85 dBA at 3 feet and that sound power levels are not to exceed 95 dBA, should be made to manufacturers for any new motors purchased. The following manufacturers have developed quiet motors:

Westinghouse Electric Corp.
Medium Motor & Gearing Div.
Buffalo, NY 14240

Reliance Electric Company
24701 Euclid Avenue
Cleveland, OH 44117

General Electric Company
Schenectady, NY 12345

The Louis Allis Company
427 E. Stewart
Milwaukee, WI 53201

TABLE 39.1

A-WEIGHTED MOTOR SOUND POWER LEVELS
(dBA re 10^{-12} watt)

Horsepower	NATIONAL ELECTRICAL MANUFACTURERS ASSOCIATION 3600 rpm DPPR	3600 rpm TEFC	1800 rpm DPPR	1800 rpm TEFC	INTERNATIONAL ELECTRO-TECHNICAL COMMISSION 3600 rpm DPPR	3600 rpm TEFC	1800 rpm DPPR	1800 rpm TEFC	A LEADING MOTOR MANUFACTURER'S DATA 3600 rpm DPPR	3600 rpm TEFC	3600 rpm QUIET LINE	1800 rpm DPPR	1800 rpm TEFC	1800 rpm QUIET LINE
1			70	74		86		80				60	63	63
1.5	76	88	70	74		91		83	65	78	78	60	63	63
3	76	91	72	79	97	100	88	87	65	78	78	60	63	63
5	80	91	72	79	97	100	88	91	65	83	83	64	68	68
7.5	80	94	76	84	97	100	88	91	68	83	83	64	68	68
10	84	94	76	84	100	103	92	91	68	86	79	68	73	73
15	84	98	80	89	100	103	92	96	72	86	79	68	73	73
20	87	98	80	89	100	103	92	96	72	94	80	74	77	77
25	87	100	83	92	102	105	94	97	77	93	84	74	82	78
30	90	100	83	92	102	105	94	97	83	93	84	74	86	78
40	90	103	86	97	104	107	97	99	83	94	88	74	86	76
50	94	103	86	97	104	107	97	99	84	94	88	74	89	76
60	94	105	89	100	106	109	100	103	84	94	84	82	89	78
75	98	105	89	100	106	109	100	103	91	94	84	82	90	78
100	96	106	92	102	106	109	100	103	91	95	88	82	101	78
125	102	107	92	102	108	112	103	106	96	100	89	85	101	91
150	102	107	94	104	108	112	103	106	96	100	89	85		91
200	105		94	104	108	112	103	106	97					91
250	105				110	114	106	109	97					

TEFC: Totally enclosed fan

DPPR: Dripproof

QUIET LINE: Change of fan blade plus other motor properties

-126-

40 - AIR NOISE

The generation of air noise results from the creation of fluctuating pressures due to turbulence and shearing stresses as high velocity gas interacts with the ambient air or solid surfaces. Radiating sources called "eddies" are formed with the high frequency noise being generated in the mixing shearing region and the lower frequency noise being generated downstream in the region of large scale turbulence.

Theoretically, as the pressure ratio between reservoir (line pressure) and ambient air is increased, the velocity of the air at the discharge nozzle increases. However, when a pressure ratio of approximately 1.9 is reached, the flow velocity through the nozzle becomes sonic, i.e., reaches the speed of sound, and further increases in reservoir pressure do not significantly increase the flow velocity. When this critical pressure ratio of 1.9 is reached, the nozzle is said to be "choked". It may be assumed, however, that at 80-100 psi, the air jets are generally in a condition of choked flow.

Based on the work of Lighthill, the overall sound power from a subsonic or sonic jet be calculated from:

$$P = \frac{K\rho A v^8}{c^5} \qquad (1)$$

where:
- A = area of jet nozzle
- ρ = density of ambient air
- v = jet flow velocity
- c = speed of sound
- K = constant of proportionality

It is clear that the velocity of the gas stream has the greatest influence on jet noise. Cutting the velocity in half would lower the sound power by 24 decibels. However, halving the area would account for a decrease of only 3 decibels. For practical calculation of jet noise, Equation (1) may be modified to the following form:

$$P = \frac{eM^5\rho V^3 A}{2} \qquad (2)$$

where:
- V = average flow velocity through nozzle
- M = Mach number of flow (V/c)
- ρ = density of ambient air
- A = nozzle area
- e = constant of proportionality of the order of 10^{-4}

For choked flow (Mach 1) conditions, the factor eM^5 is approximately 3 x 10^{-5}.

The sound power from equation (2) may be converted to decibels utilizing the relationship:

$$L_W = 10 \log \frac{P}{10^{-12}} \tag{3}$$

The frequency of peak noise level can be calculated to the first order from the following equation:

$$f_O = \frac{SV}{D} \tag{4}$$

where: S = Strouhal number (a constant) = 0.2 approx.
           for a wide range of conditions
         V = nozzle exit velocity
         D = nozzle diameter

Air discharges are observed throughout the plant and are considered a primary source of excessive sound levels in most areas. Most air discharges result from pneumatic exhaust of control systems, and these air exhausts are either continuous or cyclic in nature.

Sound levels generated by air exhaust, where concentrated air flow is not required, may be silenced by the installation of pneumatic mufflers which diffuse the air stream. The specification of an optimum pneumatic silencer for any application should be based upon:

- Sound level reduction
- Low pressure drop
- Durability
- Non-clogging features for the air contaminants present
- Economy

Mufflers should be periodically inspected for wear and effectiveness. If necessary an in-line filter system may be installed to prevent muffler clogging.

The following are manufacturers of pneumatic exhaust mufflers:

Permafilter
Div. Bonded Products, Inc.
P.O. Box 263
Wrentham, MA 02093

The Aeroacoustic Corp.
P.O. Box 65
Amityville, NY 11701

Allied Witan Company
12500 Bellaire Road
Cleveland, OH 44135

The following information should be provided to pneumatic muffler suppliers to assist in proper muffler selection:

a. Pipe thread size (N.P.T.).

b. Estimated or measured air flow (cfm).

c. Presence of air line contaminants (oil, excess moisture, etc.).

It should be pointed out that most air discharges have threaded outlets which would easily accept silencers. To insure that the silencers are not removed, they may be secured by means of welding or a set screw. Signs reminding employees that noise control devices are for their benefit may also help.

Air leaks from pneumtaic systems are often found to be major contributors to the overall noise in many plants. It is important that a program be implemented including an inspection and maintenance procedure, performed on a regular basis, to identify air leaks such as due to the following:

• Broken air hoses and cracked pipes
• Worn fittings and couplings
• Faulty valves
• Air-operated devices left on when not in use

It should also be noted that such a maintenance program would be a significant cost and energy conservation measure, as well as an important part of a noise abatement program. The air consumption and cost associated with air leakage from a sharp-edged orifice continuously at 100 psig with air at 7 cents per 1000 cubic feet is as follows:

Diameter of Opening-Inches	Air Flow, cfm	Annual Cost of Waste (8 hour shifts)	Annual Cost (continuous)
1/32	1.62	$14.19	$59.60
1/16	6.49	$56.85	$238.78
1/8	26.0	$227.76	$956.59
1/4	104.0	$911.04	$3826.37

41 - METAL-TO-METAL IMPACT NOISE

Where structures or parts are impacted with metal-to-metal surface contact, a large portion of the impact energy is converted to vibrational energy, and in turn sound. The acoustical energy generated by part impacts is directly proportional to the kinetic energy at the impact point.

$$I_a \propto K.E. = \tfrac{1}{2}mv^2$$

where: I_a = acoustical energy
K.E. = kinetic energy
m = part mass
v = part velocities

One method of reducing impact noise is to modify the system to reduce the impact velocity. Another method is to interrupt the metal-to-metal contact with a cushioning material, which serves to reduce the momentum transfer of the impacting structure.

Four types of impact surfaces may be considered, and it is found that the strongest of the materials will offer the least isolation. Selection of an optimum material must be made on an experimental basis. Three classes of materials to be considered are as follows:

1. The highest degree of impact isolation is offered by a rubber surface. It is found that rubber will wear very well in many situations, but is unacceptable in others. Rubber materials with good wear properties are:

Material	*Manufacturer*
Trellex 60	Trelleborg Rubber Company, Inc. 30700 Solon Industrial Parkway Solon, OH 44139
Armaplate and Jade Green Armabond	Goodyear Tire & Rubber Company Industrial Products Division Akron, OH 44136
Metalbak	Linatex Corporation of America Stafford Springs, CT 06076

2. Of great interest for highly stressed mechanical components are the plastics and their characteristics listed in Table 41.1. We have contacted several major manufacturers of plastics and find the following products available for impact isolation:

Product	*Manufacturer*
Lexan	General Electric 1 Plastics Avenue Pittsfield, MA 01201
Zytel ST 801	E.I. DuPont de Nemours Co. Plastics Department 170 Mount Airy Road Basking Ridge, NJ 07920

3. An impact energy absorbing foam, C3002-7, is available from:

> Doug Biron Associates
> P.O. Box 413
> Buford, GA 30518

This material may also be faced with an exterior layer of damped sheet metal to serve as a protective facing.

TABLE 41.1

TYPICAL PLASTIC MATERIALS FOR IMPACT CONTROL

A. *Nylon.* For general purpose gears, mechanical components; has good vibration damping, machines well, resists oils and solvents.

B. *Acetals.* For accurate parts, maximum fatigue life, good machineability and resistance to oils, solvents and alkalis.

C. *TFE-Fiber Filled Acetals.* For heavy duty applications, excellent wear life, creep resistant, self-lubricating, low friction, good machineability and resistance to oils and solvents.

D. *Polycarbonates.* For intermittent very high impacts (not for repeated cyclic stress), creep resistance, dimensional stability, machines well, resists acids.

E. *Fabric-Filled Phenolics.* For low cost stamped gears or parts, creep resistance, good mechanical strength, resistance to oils and solvents.

F. *Glass-Filled Phenolics.* For highest mechanical strength and temperature resistance.

G. *Glass Fibric Epoxy.* For highest electrical and mechanical properties.

H. *Polyester.* High impact.

Reference

Mazoh, M., "Modern Materials for Noise Control," Design Engineering Conference, May 10, 1972.

42 - ACOUSTICAL CURTAINS

Flexible curtains of mass-loaded vinyl have excellent sound transmission
loss properties and may be used to block airborne sound. Typical applications include:

- Complete or partial machine enclosures.

- Moveable walls to isolate noisy machine areas from
  other quieter areas.

- Localized enclosures for machine parts.

Acoustical curtains are constructed of 0.5 to 1.0 psf lead-filled vinyl and
may be installed with grommets or on sliding tracks. It is important that
the curtains extend to the floor and be sealed at the edges with Velcro fasteners for maximum sound attenuation. Installation details of enclosure
construction are given in Figure 42.1. The following manufacturers provide
acoustical curtains:

> Armstrong Cork Company
> Lancaster, PA 17604
>
> Doug Biron Associates
> P.O. Box 413
> Buford, GA 30518
>
> Ferro Corporation
> 34 Smith Street
> Norwalk, CT 06852
>
> Industrial Noise Control, Inc.
> 785 Industrial Drive
> Elmhurst, IL 60126
>
> Singer Partitions, Inc.
> 444 North Lake Shore Drive
> Chicago, IL 60611

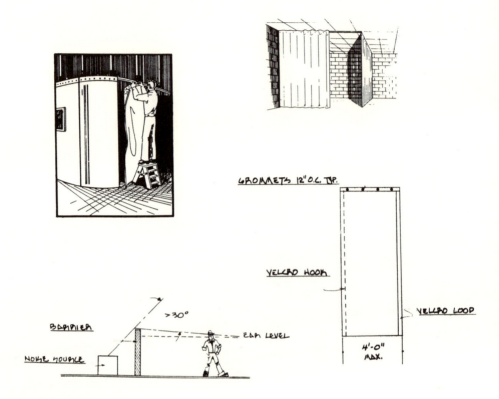

Figure 42.1. Installation guidelines for acoustical curtains.

43 - EMPLOYEE ENCLOSURES AND BARRIERS

Enclosures, partial enclosures, and barriers are often considered as measures to reduce employee noise exposure from adjacent operations. These measures may be quite effective in some instances, however, may be ineffective or actually increase an operator's noise exposure in other circumstances. Two general guidelines should govern enclosure and barrier application:

1. An employee enclosure is effective only when an employee's job tasks allow him to spend a significant portion of his workday in an enclosure.

2. An acoustical barrier is effective only when the receiver is in the direct field rather than reverberant field of a noise source.

If a four-sided enclosure is applicable, windows, ventilation, and communication equipment should be installed. This type of booth can be used where the operator does not rely on audible detection techniques for operation. This kind of enclosure is ideal for control rooms and for rest area locations where there are often employees present who are not directly involved in the area operations and are present for environmental benefits.

A three-sided, or lean-to type enclosure is applicable where only a small noise reduction is required. Such an enclosure provides safety as well as production advantages. In operations which require that the operator be able to hear the machine, this type of enclosure may reduce sound levels to an acceptable level while permitting effective audible monitoring of the operation. The enclosure may have several openings or exits for accessibility and minimization of potential entrapment hazards, and can be equipped with windows as required. The following is a list of manufacturers of commercially available enclosures:

> Keene Corporation
> 2319 Grissom Drive
> St. Louis, MO 63141
>
> Industrial Acoustics Company, Inc.
> 380 Southern Boulevard
> Bronx, NY 10454
>
> Doug Biron Associates
> P.O. Box 413
> Buford, GA 30518
>
> Eckel Industries, Inc.
> 155 Fawcett Street
> Cambridge, MA 02138
>
> Ross Engineering
> P.O. Box 751
> New Brunswick, NJ 08903

Sound Fighter Systems
1200 Mid-South Towers
Shreveport, LA 71101

An acoustical barrier, where applicable in an industrial environment, would seldom provide more than a 3-7 dBA noise reduction. To achieve this reduction, the barrier must completely block the line-of-sight path between a noise source and the observer. The barrier material must be without holes or openings, and should be of 0.5 pound per square foot minimum weight. The distance from a noise source where the reverberant sound field is equal to the direct sound field may be computed from:

$$r = 0.14\sqrt{\bar{\alpha}S}$$

where: r = distance from source, ft
$\bar{\alpha}$ = average sound absorption coefficient
S = total surface area of the interior space, ft^2

For an observer located at this distance from a noise source, a barrier may provide a 3 dBA maximum noise reduction. A barrier would be ineffective where an observer is located a greater distance from the noise source.

Machine inefficiencies, wear and malfunction can result in significant increases in noise levels. The goal of normal plant maintenance is to keep machinery in proper operating condition for efficient production. By simply expanding the existing maintenance program through noise awareness, it is possible to minimize the noise environment within the plant. If noise criteria are not assigned a level of priority, a machine with noise-producing maintenance problems, but which is still operating at 100% efficiency, would not justify maintenance attention. Because of this fact, acoustical maintenance may be met by the same opposition as its counterpart, preventive maintenance; cost savings are often hidden rather than direct.

Acoustical maintenance (AM) must be viewed as a separate discipline and an added responsibility to normal maintenance procedures. Noise control through maintenance would require an educational program for maintenance personnel to create "noise awareness" and outline engineering basics similar to those employed for preventive maintenance. This will provide maintenance personnel with the technical background necessary to cope with the unique engineering features associated with noise reduction. An effective program would involve creating a "noise awareness" among all personnel, and designating individuals whose primary duties are those related to plant-wide noise reduction. These duties may also include general energy conservation. A training program for maintenance personnel may follow the following outline:

        I.   Introduction to Noise Control
             A.   Decibels
             B.   OSHA
             C.   Sound Level Meters
             D.   Materials

       II.   Machine Design and Noise Control
             A.   Air Sources
             B.   Mechanical Sources

      III.   Application
             A.   Inspection Procedures
             B.   Guidelines for Installation

The importance of noise control related to maintenance cannot be over-emphasized. In a survey of 195 machines, it was found that over 60% were operating with a sound level at least 3 dBA louder than should be expected for a well-maintained machine. In one case, two identical machines were observed to have sound levels of 84 dBA and 107 dBA due to maintenance problems in one machine.

To organize an acoustical maintenance program, a maintenance engineer should be trained in noise control. Following this training, his duties would be as follows:

1. The engineer would perform periodic inspections to identify noise problems related to maintenance.

2. He would report noise problems to the the maintenance department to schedule for repair.

3. The engineer would specify and order any acoustical materials not in stock.

4. He would consult with maintenance personnel if technical questions should arise regarding implementation.

5. He should inspect the repairs upon completion.

6. The engineer should maintain records regarding noise control inspections and repairs.

The first task of the engineer is to document sound levels of each item of machinery when in good maintenance condition. This data serves as a baseline to identify when excessive noise is present during future surveys. Following the development of baseline data, noise survey inspections should be performed periodically (typically at one month intervals), and machine problems which cause measured sound levels to be 2-3 dBA higher than the baseline data should be identified and reported for repair. In addition, careful visual inspection should be made of each machine during each inspection. Particular attention should be given to:

- Alignments and adjustments
- Vibration and impact treatments
- Air systems
- Lubrication
- Machine dynamics
- Acoustical installations

Manufacturers of acoustical materials and the Insurance Services have adopted a system using the flame spread classification (FSC) to rate combustible and non-combustible materials. The classification system is as follows:

Class	Flame Spread (FSC)	Characteristic
1	25 or less	Non-combustible
2	26 to 75	Combustible
3	76 & above	Combustible

Underwriters Laboratories, Factory Mutual Research, and Southwest Research Laboratories test and rate materials on a client basis under ASTM E-84. The U.L. 723 tunnel test, approved as ANSI A2.5-1970, is synonymous with the E-84 test method. Other standards for fire resistance rating of wall, floor, and ceiling materials are presented in U.L. 263 and ASTM E-119. Also, fire ratings for doors and wall systems are commonly presented on a time basis.

The engineer should insure that acoustical installations meet fire code standards, or plant safety may be jeopardized, and insurance rates may be raised.

Building construction is under the jurisdiction of the National Fire Protection Association. Local codes may add extra supplemental regulations without altering the basic specification. The local inspectors play a major role in the acceptance of materials installed. Insurance rates are readjusted according to compliance with national fire codes, nature of occupancy, and the fire protection class, rated 2 to 10 (best to worst respectively). Rates may be affected up to 100%.

In the case of enclosures with foam lining, the running footage of the building is calculated. The enclosure area is considered as a partitioned wall, and its running footage is also calculated. If the footage of the enclosure is less than 25% of the overall, there is no rate adjustment. If the footage is 25% to 50%, a 10% increase can be expected. If the percentage is 50% or more, the increase would be 20%; if it is 100% or over, there is a 40% increase. Where footage equals or exceeds 200%, a 100% rate increase is applied.

Generally, use of foam in a building, especially vertical applications, would change a non-combustible metal building rating (NC-2) to a combustible frame building rating (wood). A similar condition would occur where combustible spray-on acoustical treatment is utilized. The rating would change from a metal building to a frame building. For example, in a protection class of 3, the base rate of $.125 for a metal building would increase to $1.90, the rate for a frame building. Ceiling-hung combustible baffles or ceiling tile may increase rates 50% to 100%, depending on the amount of concealed area. Trowel-on dampening, a material non yet investigated, would fall under a rate schedule for hazardous conditions. Depending on the severity of the conditions, the rates could increase from 50 to 400%.

Generally, changes from a metal, non-combustible building rating to a masonry rating can increase rates on an average of 100 to 150%, and further, changes to a frame building rating (wood, combustible) can increase the rates an average of 150 to 200%.

A letter requesting a list of approved materials and/or fire ratings was sent to over 200 manufacturers of acoustical materials. Sixty-eight percent of those responding did not present fire ratings of any type in the consumer-available literature. Table 45.1 presents a list of all acoustical manufacturers in the United States known to provide fire ratings for their materials and systems.

In some cases acoustical materials have not been tested due to the fact that their classification is governed by the system for which they are used. For this reason, it is difficult to rate some acoustical materials, such as trowel-on dampening, since they are a part of an integral system, and the industrial machines for which they are used have not been considered for classification at all. There are acoustical systems, such as composite panels of perforated sheet metal, foam or glass bonded to sheet metal, which are commonly used, and still have not been listed.

TABLE 45.1

MANUFACTURERS OF FIRE-RATED
NOISE REDUCTION PRODUCTS

MANUFACTURER	PLYWOOD	GLASS FIBER	ACOUSTICAL CEILING SYSTEMS	DAMPING	SPRAY ABSORPTION	DOORS	SILENCERS/MUFFLERS	GYPSUM BOARD/PARTICLE BOARD	PLASTICS	LEAD	FOAM	CONCRETE BLOCK/CERAMICS	LEAD-LOADED VINYL/LOADED VINYL	METALS	ACOUSTICAL WALL SYSTEMS	ACOUSTICAL PANELS
Aeroacoustic Corporation 1465 Strong Avenue Copiague, NY 11726 (516) 226-4433		•			•											
Air-O-Plastik Corporation Asia Place Carlstadt, NJ 07072 (201) 935-0500		•							•							
Alpro Acoustics Division Structural Systems Corporation P.O. Box 30460 New Orleans, LA 70190 (504) 522-8656		•													•	•
American Smelting & Refining Co. 150 St. Charles Street Newark, NJ 07101 (201) 589-0500										•						
Arrow Sintered Products Company 7650 Industrial Drive Forest Park, IL 60130 (312) 921-7054							•									
Brunswick Corporation 1 Brunswick Plaza Skokie, IL 60076 (312) 982-6000														•		

TABLE 45.1 (Continued)

MANUFACTURER	PLYWOOD	GLASS FIBER	ACOUSTICAL CEILING SYSTEMS	DAMPING	SPRAY ABSORPTION	DOORS	SILENCERS/MUFFLERS	GYPSUM BOARD/PARTICLE BOARD	PLASTICS	LEAD	FOAM	CONCRETE BLOCK/CERAMICS	LEAD-LOADED VINYL/LOADED VINYL	METALS	ACOUSTICAL WALL SYSTEMS	ACOUSTICAL PANELS
Certain-Teed Products Corporation CSG Group Valley Forge, PA 19481 (215) 687-5500		•														
Conwed Corporation 2200 Highcrest Road Saint Paul, MN 55113 (612) 645-6699			•													
Doug Biron Associates, Inc. P.O. Box 413 Buford, GA 30518 (404) 945-2929		•	•							•	•		•		•	
Ferro Corporation 34 Smith Street Norwalk, CT 06852 (203) 853-2123												•	•			
Globe Industries, Inc. 2638 E. 126th Street Chicago, IL 60633 (312) 646-1300														•		
Gypsum Association 201 North Wells Street Chicago, IL 60606 (312) 491-1744	•							•								

-142-

TABLE 45.1 (Continued)

MANUFACTURER	PLYWOOD	GLASS FIBER	ACOUSTICAL CEILING SYSTEMS	DAMPING	SPRAY ABSORPTION	DOORS	SILENCERS/MUFFLERS	GYPSUM BOARD/PARTICLE BOARD	PLASTICS	LEAD	FOAM	CONCRETE BLOCK/CERAMICS	LEAD-LOADED VINYL/LOADED VINYL	METALS	ACOUSTICAL WALL SYSTEMS	ACOUSTICAL PANELS
The Harrington & King Perforating Co. 5655 Fillmore Street Chicago, IL 60644 (312) 626-1800														●		
Holcomb & Hoke Manufacturing Co., Inc. P.O. Box A-33900 Indianapolis, IN 46203 (317) 784-2444						●										
Koppers Company, Inc. Pittsburgh, PA 15219 (412) 319-3300	●															
Metal Building Interior Products Co. Lakeview Center 1176 E. 38th Street Cleveland, OH 44114 (216) 431-6040		●	●													●
National Cellulose Corporation 12315 Robin Boulevard Houston, TX 77045 (713) 433-6761					●											
Nichols Dynamics, Inc. 740 Main Street Waltham, MA 02154 (617) 891-7707		●												●		

-143-

TABLE 45.1 (Continued)

MANUFACTURER	PLYWOOD	GLASS FIBER	ACOUSTICAL CEILING SYSTEMS	DAMPING	SPRAY ABSORPTION	DOORS	SILENCERS/MUFFLERS	GYPSUM BOARD/PARTICLE BOARD	PLASTICS	LEAD	FOAM	CONCRETE BLOCK/CERAMICS	LEAD-LOADED VINYL/LOADED VINYL	METALS	ACOUSTICAL WALL SYSTEMS	ACOUSTICAL PANELS
Pittsburgh Corning Corporation Geocoustic Systems 800 Presque Isle Drive Pittsburgh, PA 15239 (412) 261-2900			•													
The Proudfoot Company, Inc. P.O. Box 9 Creenwich, CT 06830 (203) 869-9031												•		•		
Scott Paper Company Foam Division 1500 E. 2nd Street Chester, PA 19013 (215) 876-2551											•					
Singer Partitions, Inc. 444 N. Lake Shore Drive Chicago, IL 60611 (312) 527-3670				•												
Specialty Composites Delaware Industrial Park Newark, DE 19713 (302) 738-6800				•								•		•		
Stark Ceramics, Inc. P.O. Box 8880 Canton, OH 44711 (216) 488-1211												•				

TABLE 45.1 (Continued)

MANUFACTURER	PLYWOOD	GLASS FIBER	ACOUSTICAL CEILING SYSTEMS	DAMPING	SPRAY ABSORPTION	DOORS	SILENCERS/MUFFLERS	GYPSUM BOARD/PARTICLE BOARD	PLASTICS	LEAD	FOAM	CONCRETE BLOCK/CERAMICS	LEAD-LOADED VINYL/LOADED VINYL	METALS	ACOUSTICAL WALL SYSTEMS	ACOUSTICAL PANELS
Starco 1515 Fairview Avenue St. Louis MO 63132 (314) 429-5650		•						•	•							
U.S. Plywood Div. of Champion International 777 Third Avenue, New York, NY 10017 (212) 895-8000	•						•									
Veneered Metals, Inc. P.O. Box 327 Edison, NY 08817 (201) 549-3800				•										•		

46 - LONG TERM NOISE ABATEMENT

In the specification and purchase of all new items of equipment, requirements for reduced product noise should be included as part of the specification. One model noise control specification is as follows:

General

(1.1) The manufacturer shall submit sound measurements for supplied
equipment in accordance with this specification. In addition,
where the sound measurements exceed the values stated below,
the manufacturer shall advise on silencing provisions and
additional costs to meet this standard.

Measurement

(2.1) The manufacturer shall be responsible for the supplied equipment,
including any subassemblies.

(2.2) The manufacturer shall test the equipment as follows:

Item	Applicable Standard
Duct & Fan Systems	AMCA 300-67 Rest Code
Rotating Electric Machinery	IEEE 85
Pneumatic Equipment	ANSI S5.1
Machine Tools	NMTA Noise Measurement Techniques

Where no acceptable sound test exists or where manufacturer's
standard is different from the above, the manufacturer shall
submit the method of testing.

Submitting Data

(3.1) The manufacturer shall submit octave sound power data for equipment. In addition, the following information is required:

Item	Information Required
Fans	a) Total sound power level at fan casing
	b) Inlet sound power level
	c) Outlet sound power level

(Item)	(Information Required)
	d) Locations where sound measurements were taken
Rotating Machinery and Miscellaneous Devices	a) Total sound power level at the equipment casing
	b) Locations where sound measurements were taken

(3.2) If the source is highly directional, such as a cooling tower, measurements shall be submitted at several locations.

Specified Levels

(4.1) If manufacturers's data submitted in Items 3.1 and 3.2 exceeds 85 dBA, the quotation shall include the additional cost and silencing provisions necessary to meet this value.

Manufacturer's Responsibility

(5.1) It is the manufacturer's responsibility to engage an independent consultant, as required, in order to meet the noise requirements stated in this specification.

(5.2) The manufacturer shall not ship any equipment which exceeds the manufacturer's values promulgated in this specification without the purchaser's written authorization.

(5.3) If the manufacturer must perform tests at the purchaser's facility, it shall be stated in the quotation.